p 146 - "neither is
to have been a
what the signif
mere."
157-158 hint of a tragic sense
(but not in universities.

Democracy and the Novel

Democracy and the Novel

POPULAR RESISTANCE TO CLASSIC AMERICAN WRITERS

Henry Nash Smith

OXFORD UNIVERSITY PRESS
Oxford New York Toronto Melbourne

Oxford University Press
Oxford   London   Glasgow
New York   Toronto   Melbourne   Wellington
Nairobi   Dar es Salaam   Cape Town
Kuala Lumpur   Singapore   Jakarta   Hong Kong   Tokyo
Delhi   Bombay   Calcutta   Madras   Karachi

First published by Oxford University Press, New York, 1978

First issued as an Oxford University Press paperback, 1981

Library of Congress Cataloging in Publication Data

Smith, Henry Nash.
  Democracy and the novel.

  Includes bibliographical references and index.
  1. American fiction—19th century—History and
criticism.   2. United States—Popular culture.
3. Literature and society.   I. Title.
PS377.S55   813′.03   78-1290
ISBN 0-19-502397-8
ISBN 0-19-502896-1 pbk.

Printed in the United States of America

For Harriet, Janet, and Mayne

Acknowledgments

The ideas in this book have been worked out over a period of years in seminars that I have had the pleasure of conducting at the University of California (Berkeley) and, as a visitor, at Washington University. The result is therefore in an important sense a joint enterprise of perhaps a hundred young scholars, whose substantial contributions I wish to acknowledge with lively gratitude even though it is not feasible to list their names here. I can, however, thank individually five students who have helped me as research assistants: Lorne M. Fienberg, Jay E. Gillette, Robert H. Hirst, Mary S. Holland, and Janet Irwin Smith.

The older colleagues on the faculties of my own and other universities who have offered criticism and advice are also too numerous to mention, but I owe a special debt to Leo Marx and Henry May, both of whom disagree with me (from radically different perspectives) on some issues of theology and political and literary theory, but whose challenges have been indispensable to the development of my own thought. Sheldon Meyer has also helped me more than editors are expected to, not least by his Olympian serenity concerning missed deadlines.

It has been a luxury to be able to depend on the powers of three typists—Anita Lynn, Linda Schneider, and Marie Herold—to decipher pages of typescript covered with layers of handwritten revisions.

Two chapters of this book are revised versions of articles that have appeared in journals. I wish to thank the editors of the *Yale Review* and *Prospects,* respectively, for permission to reprint "The Madness of Ahab" and "A Textbook of the Genteel Tradition: Henry Ward Beecher's *Norwood.*"

Finally, I am glad to have this opportunity to express my thanks for financial support from the Senate Committee on Research of the Berkeley Division of the University of California, the John Simon Guggenheim Memorial Foundation, and the Woodrow Wilson International Center for Scholars. The Wilson Center gave me the additional intellectual stimulus of daily association over a long period with a select community of scholars.

Berkeley
January 1978   H.N.S.

Contents

Democracy and the Novel

The Issues

For more than a generation, under the influence of F. O. Matthiessen's *American Renaissance* (1941), we have recognized the years 1850–1855 as marking a culmination in the history of our literature. In fiction, especially, Matthiessen made us aware that the publication of *The Scarlet Letter* in 1850 and of *Moby-Dick* in 1851 represented an "extraordinarily concentrated moment of expression." Proposing to do justice both to Vernon L. Parrington's political concern with "what our fathers thought" and to the esthetic emphasis of the New Criticism, he asserted that "the one common denominator" of the writers he dealt with was "their devotion to the possibilities of democracy," yet he also declared that his duty as a critic was "to be preoccupied with form."[1] Almost four decades later, Matthiessen's program remains as an ideal and a challenge. We gratefully take his work for granted as he took Parrington's.[2] At the same time, we can recognize that his conception of what is meant by a writer's concern with democracy was almost entirely limited to the ideas and attitudes expressed in the man's work. Matthiessen was much less interested in the relation of literary masterpieces to social processes or to non-literary aspects of the culture. The suggestive but brief comments on genre painting and backwoods humor in *American Renaissance* come almost as afterthoughts and have a somewhat perfunctory air that contrasts markedly, for instance, with the intensity of the analyses of *Moby-Dick* or *Walden*. And Matthiessen explicitly relegates to "the sociologist and . . . the historian of our taste" the popular culture embodied in the kind of fiction actually preferred by the largest number of people.

The esthetic value of the American classics has now been established so firmly that we can examine their relation to the cultural setting without danger of losing sight of what makes them masterpieces. In particular, the concern with popular culture that has burgeoned during the past decade has directed attention in much greater detail both to similarities and to differences between serious literature and the reading matter produced for sale in a mass market. Since no artist works in a vacuum, there is always something to be learned from attempting to place ourselves within a writer's intellectual and cultural horizon. That of nineteenth-century American novelists was dominated by the towering eminences of Scott and Dickens, but it also included now forgotten contemporaries and immediate predecessors who are of no esthetic consequence but are highly useful as indices to the attitudes prevalent among readers generally. Unlike Emerson, Thoreau, and Whitman, the other writers on Matthiessen's list, Hawthorne and Melville depended on their writing for a livelihood, and for this reason were obliged to pay close attention to the preferences of the book-buying public. The same was true of William Dean Howells, Mark Twain, and Henry James, the three later writers I intend to consider. Some of them enjoyed longer or shorter periods of considerable popularity (Melville, for example, with his first stories of adventure in the South Seas, James with the mild *succès de scandale* of *Daisy Miller*), and of course Mark Twain was a conspicuous best-seller; but none of these writers was able to establish a durable rapport with an audience that allowed him to develop his full potentialities. Hawthorne, Melville, and James each suffered a drastic decline in sales in mid-career. During the last thirty years of Howells's life he virtually abandoned his struggle against the conventions that bound the novel to the prayer-wheel of the genteel love story; and most of Mark Twain's critics refused to acknowledge that he could be anything except a funny man. Despite his emphatic repudiation of the prevalent American optimism in *A Connecticut Yankee* and the posthumous *The Mysterious Stranger*, these books were received as, respectively, a glorification of American technology and democracy and a Christmas fairy-tale for children. I have long been fascinated by the solidity, the durability, the imperviousness of the secular faith or ideology lying at the base of American popular culture; the present book grows out of my gradual realization that each of the major novelists examined here came into collision with it. There is a sense in which I have not transcended Matthiessen's distinction after all, for much of the time I am investigating social or cultural history rather than literature. On the other hand, I think I have also learned something new about several masterpieces of our fiction. Let me indicate briefly the conclusions I have reached.

My discussion of Hawthorne focuses on his challenge to the prevailing common-sense philosophy that took for granted the existence of a stable extra-mental universe completely independent of consciousness yet accurately knowable by all normal observers. The chapter on Melville is similarly restricted in subject; it deals with only one aspect of a single character, the madness of Ahab. Since this spiritual state, at once disease and heroic exaltation, marks the ultimate degree of alienation from society, the writer's barely controlled impulse to identify himself with Ahab inevitably antagonized his readers. By the outbreak of the Civil War both Hawthorne and Melville had in effect been rejected by a reading public devoted to what Hawthorne ironically called "a common-place prosperity, in broad and simple daylight."[3]

The extreme confusion that prevailed in the domain of letters at the end of the Civil War has led literary historians to postulate a virtually complete breach of continuity between pre-War "romanticism" and post-War "realism." But I think a few lines of connection can be made out, and I use Henry Ward Beecher's novel *Norwood* as a convenient means of tracing some of them. Beecher was a leading exponent of the liberal Christianity that provided the intellectual framework for the dominant American culture of the 1870s and 1880s. The most obviously representative man of letters during these years was William Dean Howells. He is a pivotal figure in this study because he was the only one of my five novelists who deliberately and persistently tried to affirm rather than challenge the prevailing system of values. Yet devotees of "idealism" (that is, the formulaic domestic fiction of the 1850s and 1860s) denounced the doctrine of realism in literature as alien and anti-Christian, and in the end he was forced to abandon the aspects of his program that were seriously disturbing to much of his audience.

The discussion of Mark Twain draws on George Santayana to argue that because Mark Twain was a humorist rather than a novelist, he could use American vernacular speech as a way out of the stylistic impasse in which novelists were trapped. Santayana, of course, was not concerned with the technique of the novel, but his assumption that there is a close connection between philosophical ideas and literary practice is precisely the assumption I make in the present study. What he means by the genteel tradition approximates what I mean by the value system of the dominant or popular culture.[4] Finally, I am glad to be able to cite Santayana in support of my bracketing of Henry James with Mark Twain as another of the very few writers who had freed themselves from the control of "the polite and conventional American mind" (p. 52), although of course James's method—"turning the genteel American tradition . . . into a subject matter for analysis" (p. 54)—was entirely different from Mark Twain's.

The question might then be raised why a book is needed to say in many words what Santayana says elegantly in so few. A part of the answer lies in the extraordinary brevity of his observations; but a greater part lies in the fact that Santayana is interested in nothing American before the Civil War except Emerson, whereas one of my principal concerns has been to construct a historical bridge over the chasm that most accounts of American literature leave between the pre-War and post-War eras.

As I have indicated, my starting-point is the 1850s. I might take my text from Matthiessen's colleague and friend, Perry Miller, who placed on record a dissent from the idea of a triumphal American Renaissance. At the very moment when Hawthorne and Melville reached the height of their achievement as artists, Miller points out, they were "being crushed before the juggernaut of the novel," with the result that they "went down"—although with "the ensign of the Romantic still flying." Thus, these men "do not inaugurate a 'renaissance' in American literature; . . . they pronounce a funeral oration on the dreams of their youth, they intone an elegy of disappointment."[5]

Miller's phrase "the juggernaut of the novel" calls attention to a startling event in the history of American publishing. Between 1850 and 1855, several books by previously unknown writers suddenly attained unprecedentedly large sales. In an often-quoted letter, Hawthorne described the writers of these best-sellers as "a d----d mob of scribbling women."[6] The one example he named was Maria Cummins's *The Lamplighter*, published in 1854. *Uncle Tom's Cabin* (1852) may also have been in his mind, although the bearing of this champion best-seller on the controversy over slavery makes it such a special case that it should probably be left out of account in considering the troubles of Hawthorne and Melville. More pertinent was the publication in 1850 of Susan B. Warner's *The Wide, Wide World*, which astonished publishers and critics alike because it revealed the existence of a previously unsuspected audience for a special kind of fiction. This new sub-category of the sentimental or domestic novel, to which *The Lamplighter* also belonged, had for its central character a pre-adolescent girl whose experiences (mostly painful) up to the age of nubility at eighteen or so provided occasion for extraordinary quantities of tears and prayers.

Whereas Cooper's most successful books had sold fewer than ten thousand copies within a year of publication, and Hawthorne had been gratified when *The Scarlet Letter* sold five thousand copies in the first six months, *The Wide, Wide World* went into its thirteenth printing within two years, was eventually reprinted sixty-seven times, and by the end of the nineteenth century had sold more than a half-million copies in the United States in addition to long-lasting popularity in Great Britain. *The Lamplighter* sold forty thousand copies in the first eight weeks and in the long run may have outstripped *Wide World* in sales.[7] Other women novelists such as Mary Jane Holmes (*Tempest and Sunshine*), Caroline Lee Hentz (*The Planter's Northern Bride*), and Augusta Jane Evans (*Beulah*) also had large although not precisely recorded sales during the decade. Miller implies that such novels drew readers away from Hawthorne and Melville—that is, that the women writers were competing with them for the same market. Hawthorne himself apparently held this opinion. In the letter in which he complained that the scribbling women's books were selling "by the 100,000," he declared: "I should have no chance of success while the public taste is occupied with their trash—and I should be ashamed of myself if I did succeed." He may have been partially correct, but the sales of the new women writers were so much larger than those of any earlier American novelist that new groups of readers must have been involved. It is impossible now to determine just what did happen to the market in the early 1850s.

In the early decades of the century, as William Charvat demonstrated many years ago in *The Origins of American Critical Thought: 1810-1835*, it was assumed that the American writer addressed a relatively homogeneous reading public with tastes defined by clergymen and lawyers residing in cities of the Eastern seaboard, together with a few landed gentlemen.[8] A single standard was supported by a faculty of taste believed to be universally valid. But during the thirty years preceding the outbreak of the Civil War, the dominance of the older patrician culture was challenged and its masculine and aristocratic values were supplanted by the leveling influences exerted by rapid increases in population and wealth, by the spread of free public schools, by the evangelical movement, and especially by the cultural influence of women, who for the first time were gaining enough leisure to have time to read, and enough education to enjoy and produce books.[9]

Even though the new audience was still only a tiny fraction of the total population (some 23 millions in 1850 and 31 millions in 1860), its expansion exerted strong pressures on writers by placing an unmistakable premium on certain fictional stereotypes and formulas. We need labels for the distinction that Hawthorne had in mind when he called *The Lamplighter* "trash," and the additional distinction between the work of the new women novelists and yet another kind of fiction that proliferated during the 1840s—crude adventure stories represented by Tom Sawyer's favorite, *The Black Avenger of the Spanish Main; or, The Fiend of Blood. A Thrilling Story of Buccaneer Times*, by "Ned Buntline" (E.Z.C. Judson), published in 1847. For want of a more elegant terminology I shall call the sentimental fiction of Warner and Cummins, together with the system of values embodied in it, "middlebrow," and Ned Buntline's work "lowbrow." In such a scheme, Hawthorne and Melville evidently must be categorized as "highbrow."[10] Although not enough evidence has been accumulated to support confident statements about the sizes of these

segments of the mid-nineteenth century reading public, it appears
that the total sales of lowbrow fiction were the largest—especially
after the Beadle & Adams dime novels began to appear in 1860; but a
few best-selling middlebrow titles far outstripped any individual
lowbrow items in circulation.

Yet another area of obscurity is the relation of what I am calling
"brow levels" to social and economic classes. Along with Tom
Sawyer and his fondness for Ned Buntline must be considered the
nightwatchman on the steamboat *Paul Jones,* described in Chapter 5
of *Life on the Mississippi,* who has consumed so much "wildcat
literature" that his mind is crammed with "bloodshed and . . .
hair-breadth escapes." The implication is that such fiction appealed
to boys and relatively ignorant men. But it is far from clear that these
two groups constituted an American equivalent for the urban
"working classes" that Louis James identifies in early Victorian
England.[11] The notorious vagueness of class lines in the United States
precludes any close linkage between brow levels and the actual
social structure. Nevertheless, I shall attempt to show that from
the 1840s onward the differentiation of literary tastes can be traced
in the choice of subject matter, style, and other features of novels
produced by writers seeking to please the new audience (or audiences). Contemporary reviewers show greater and greater awareness
of differences in types of fiction that almost resemble distinct
genres.

The nineteenth-century novelists whom we now value resisted
the demands of the new middlebrow audience, yet without exception
their work was visibly influenced by the struggle—sometimes (as in
Melville's case) only slightly, sometimes (as in Howells's) to a
considerable, even decisive extent. The accommodation of writers to
the audience took many different forms. Hawthorne was diffident
toward his potential audience from the beginning. He discusses his
difficulty in communicating with an extensive public as early as 1844,
in the odd burlesque preface to "Rappaccini's Daughter." Adopting a
tone of awkward jocularity, he writes of himself in the third person
under the nonce pseudonym "M. de l'Aubépine" (the French word for
"hawthorne"). He apologizes for the lack of "human warmth" in his
stories, and blames this trait for the small circulation they have in
comparison with the popularity of Eugène Sue. Aubépine, he says,

"seems to occupy an unfortunate position between the Transcendentalists . . . and the great body of pen-and-ink men who address
the intellect and sympathies of the multitude."[12] In 1850 Hawthorne
had suggested to his publisher that the words "The Scarlet Letter" be
printed in red ink on the title page of the book (a suggestion which
was followed). "I am not quite sure about the good taste of so doing,"
he continued, "but it would certainly be piquant and appropriate—
and, I think, attractive to the great gull whom we are endeavoring to
circumvent."[13]

Melville cherished in a much more developed form the notion
that author and publisher were engaged in a conspiracy to deceive the
public. In his often quoted review of *Mosses from an Old Manse*
(1850), he maintains that Hawthorne, like Shakespeare, conceals
beneath a bland surface a "blackness" of spiritual truth which only
a few readers will perceive. Indeed, ". . . at the bottom of their
natures," declares Melville, "men like Hawthorne, in many things,
deem the plaudits of the public such strong presumptive evidence of
mediocrity in the object of them, that it would in some degree render
them doubtful of their own powers, did they hear much and vo-
ciferous braying concerning them in the public pastures." Some of
Hawthorne's sketches are in fact "directly calculated to deceive—
egregiously deceive—the superficial skimmer of pages."[14]

Melville's interpretation, although overstated, has some support
in the deprecatory tone of "The Custom House," the narrative preface
to *The Scarlet Letter*. Hawthorne writes, for example, that "when he
casts his leaves forth upon the wind, the author addresses, not the
many who will fling aside his volume, or never take it up, but the few
who will understand him, better than most of his schoolmates and
lifemates."[15] But Melville's defiance of prevalent attitudes was much
more violent than this. His response to general disapproval of "the
allegory and drama" in *Mardi* and *Moby-Dick* was to plan his next
book, *Pierre*, as an elaborate hoax. He undertook to produce a novel
that on the surface would conform to the conventions of mass fiction
as these were represented in weekly "story papers," yet would embody
an undercurrent of subversive implication repudiating the basic
articles of the dominant value system. He believed he could write the
book so that this latent meaning would be perceived by the select few
capable of responding to the thoughts of a man of genius, but would

not be recognized by "the tribe of 'general readers'" who could not bear the appalling truth. As Charvat describes Melville's plan:

> The artist could be as profound as he wished without being resented if he concealed his profundities under a pleasant or sensational narrative surface through which the reader looking for mere diversion could not penetrate. Thus, greatness could be achieved in the public art of fiction in the nineteenth century, as it had been in Shakespeare's theater.[16]

The consequence of this plan, according to Charvat's highly plausible interpretation, is that the protagonist of *Pierre*, presented at the outset as an aristocratic country youth, is suddenly revealed halfway through the book as a writer, "an intellectual, and a full-blown example of the 'misunderstood Genius' . . ." This hero is portrayed at work in New York City, where he is crucified by publishers and critics. At the same time, Melville goes out of his way to ridicule and denounce the sentimental fiction that is the delight of the mass audience. The result was that the "world repaid Melville in kind for these compliments, and his reputation never recovered [during his lifetime] from the attack on *Pierre*."[17]

Whereas Hawthorne tended to believe he was competing with the scribbling women for the favor of a single homogeneous reading public, Melville assumed that the American audience for fiction was divided into two distinct segments—the undiscriminating mass of what I have called middlebrow readers, and an intellectual and moral elite having approximately the taste we ascribe to the highbrow audience in our own day. But Melville failed to carry through the scheme he had devised for taking advantage of this supposed segmentation of the reading public. In *Pierre*, as Charvat observes, "the plot, which a *Ledger* writer [i.e., a contributor to Robert Bonner's highly successful weekly story paper] would have twisted into a happy ending at the last moment, is allowed to follow its natural course to a catastrophe which kills off the good and bad alike" (p.53). The black bitter doctrine that only a madman would try to live in this world according to moral principles surfaces in the latter part of the book. Since Melville's scheme of deception was not submitted to the test, there is no way of knowing whether any considerable number of enlightened readers such as he hoped for actually existed in the 1850s. *Pierre* was universally denounced in the press, and its reception shut

Melville off "from any further serious consideration by contemporary critics" (p. 51).

The predicament of the American writer as Melville represents it in *Pierre* has two distinct aspects. One concerns doctrine: the problems of truth about the cosmos, about morality, about society that he had encountered as soon as he had realized that writing could be more than a mere spinning of yarns about exotic adventure. His struggles with these problems were in large part personal. But his rapid development as a writer during the four years between *Omoo* and *Moby-Dick*, especially his discovery of Hawthorne while he was writing *Moby-Dick*, had plunged him into a difficulty that was technical rather than substantive and doctrinal: the question of whether a novel could appropriately concern itself with ultimate social and intellectual issues, or whether it was by definition required to be simply what the English call "a good read," a recreation rather than a serious occupation.

Conventional opinion was unanimous on this point throughout the Anglo-American world of letters. The successful English novelist Margaret Oliphant set forth the accepted doctrine clearly and firmly in some remarks about Hawthorne published in *Blackwood's Magazine* in 1855. She declared that Hawthorne misconceives the character of the novel. He has set himself to study the "inner nature of other people" and has addressed himself to "an intellectual audience," whereas the "novelist's true audience is the common people—the people of ordinary comprehension and everyday sympathies, whatever their rank may be."[18] The key word here is of course "intellectual." It accuses Hawthorne of straining the minds of his readers by compelling them to struggle with baffling problems. Mrs. Oliphant's attack is directed precisely at the achievement to which the American writer had devoted his entire career: his establishment of a realm of imaginative truth in fiction, a truth of romance distinct from the mere imitation of outer reality. Popular culture in both Britain and the United States, while accepting art (in moderate quantities) as entertainment or an ornament for moral instruction, rejected the notion that it might embody a truth not accessible through the ordinary processes of observation and interpretation.[19]

The system of values that Hawthorne was rejecting was in large part common to the two societies. But the American version incorporated a powerful element of naïve nationalism. George Santayana described the accepted articles of this faith with gentle but devastating irony: "The world . . . was a safe place, watched over by a kindly God, who exacted nothing but cheerfulness and good-will from his children; and the American flag was a sort of rainbow in the sky, promising that all storms were over. Or if storms came, such as the Civil War, they would not be harder to weather than was necessary to test the national spirit and raise it to a new efficiency."[20] The political creed dominant on this side of the Atlantic held that the United States was the last best hope of humanity: the repository of freedom, the standard-bearer of progress, the land of opportunity. Because the nation had already experienced its republican and democratic revolution, no improvement was conceivable in our form of government as it had been established in the Federal Constitution (except, of course, for the abolition of slavery, but this topic was so uncomfortable in the 1850s that it was generally avoided outside explicit partisan controversy). The majority of Americans thought of themselves as Christians, Protestant Christians in fact, and they continued to cling to this label even when the content of their religious faith had become little more than a blur of good intentions. They were convinced that religion was necessary to keep institutions in place, maintain public order, and preserve the sanctity of private property.[21] David Levin documents the widespread assumption that "History was the unfolding of a vast Providential plan."[22] The authors of the many handbooks offering advice to young men and women "took for granted a distinctly stratified society" but at the same time declared that hard work and clean living would infallibly lead to economic success and rising in the world. A man's "secular fortune was in some sense an evidence of his religious estate . . . "; indeed, public opinion tended to identify the two.[23] The strong commitment to national progress and individual success required the suppression of any hint that failure might occur except as punishment for wrongdoing. Above all, introspection was discouraged, as if it were feared that probing the inner life of ordinary men and women might bring to the surface unthinkable perversities.

Although American culture was based on a realistic metaphysics, it embraced an idealistic theory of art (especially of literature).[24] According to this doctrine, the function of art was to present images of beauty and nobility in order to inspire emulation. The writer was expected to offer readers opportunities to identify themselves with virtuous and attractive characters. Tragedy could be accommodated (in theory at least) by regarding evil and wickedness as the necessary condition for the display of moral beauty by the hero or heroine, but in practice the mode of ideality could not deal with tragedy and instead tended to foster melodrama.

It is therefore easy to understand that Hawthorne's concern with "the truth of the human heart" in his psychological romances[25] was deeply disturbing to people of ordinary comprehension and everyday sympathies, whether British or American. Mrs. Oliphant, likening "the new science which is called 'anatomy of character'" to actual dissection, picks up a cluster of images that would appear again and again later in the controversy over literary realism. The analogy generates for her as it would for later critics fantasies of body-snatching and midnight orgies of sadism in a hidden laboratory. References to "the spiritual dissecting-knife" and to a surgeon's pleasure in exhibiting his "pet 'cases'" call up such further clinical images as "a suppressed, secret, feverish excitement" observable in *The Scarlet Letter*—not "the glow of natural life, but the hectic of disease which burns upon the cheeks of its actors." In *The House of the Seven Gables*, continues Mrs. Oliphant, Clifford Pyncheon owes his existence to "the spiritual anatomist whose business it is to 'study' his neighbours"; and *The Blithedale Romance* has "still less of natural character, and more of a diseased and morbid conventional life."[26]

No reviewer in this country would have agreed with Mrs. Oliphant's complacent remark that Hawthorne's "unwholesome fascination" simply reflected attitudes prevalent among the nervous Americans; and few, if any, American critics could have achieved the tone of self-assurance with which she proclaimed the dogmas of philistinism. Nevertheless, when the fantasy life of the American Common Man and especially the Common Woman became available for inspection in the fiction preferred by the suddenly expanded reading public, it revealed that both Hawthorne and Melville, in setting out to explore the dark underside of the psyche, had been moving in a direction directly counter to that of the popular culture. Charvat, commenting on the fact that Melville's friend Evert Duy-

still undefined

ckinck felt "uncomfortable" in reading *Moby-Dick*, observes that "fiction was beginning to be, like poetry, a potential means of self-exploration for the writer. This is a function of literature which the common reader has never accepted unless it is so disguised that he need not contend with it."[27] Fiction conceived as a penetration of the unconscious seemed threatening to the newly articulate middle class, which craved not challenge but reassurance. The unstated and probably unrecognized common purpose of the best-selling novels of the 1850s was to relieve the anxieties aroused by rapid upward mobility, especially the fear of failure, and to provide assurance that the universe is managed for man's benefit.

At the deepest level of tacit assumption this popular fiction also offered the warning that is stated explicitly by the narrative voice addressing the "indulgent reader" in the last paragraph of Mrs. Southworth's novel *Retribution* (1849):

> *Divine* retribution belongs to Eternity, and is distant and vague. *Human* retribution is uncertain, depending upon discovery and other fortuitous circumstances; but *Moral* retribution is as sure as life, as sure as death, as sure as the *sin* out of whose bosom it springs, as natural as the pain that follows the contact of fire. I have tried to show you how from the sin, domestic infidelity and treachery, sprung inevitably the punishment, domestic distrust and wretchedness. Human and Legal retribution we may elude by concealment; Divine retribution we may avert by a timely repentance; but Moral retribution we must *suffer;* and that, not by the arbitrary sentence of a despot, but by the natural action of an equitable law, old as Eternity, immutable as God.[28]

I need hardly say that the handful of essays brought together here does not remotely approach either the majestic scope of the three-volume work that Parrington left unfinished at his death in 1929, or the depth of Matthiessen's definitive treatment of his five-year period. As is perhaps appropriate in the 1970s, my effort resembles instead a guerilla campaign employing the hit-and-run tactics of forces that have neither the fire power required in a full-scale engagement nor the logistic support necessary for a long-sustained operation.

Hawthorne:
THE POLITICS OF ROMANCE

Readers of Hawthorne's *The Scarlet Letter* know that the narrator of the introductory section called "The Custom House" is involved in American party politics in the crudest, most literal sense. His experiences are much like Hawthorne's own. With wry humor he recounts how as a new Surveyor of the Salem Custom House appointed by the Democrats he allowed certain white-haired veteran officers appointed by the Whigs to continue dozing the forenoons away "with their chairs tilted back against the wall," only to find that three years later, with the advent of Zachary Taylor's Whig administration, his own head "was the first that fell!"[1] This narrator, however, does not show the slightest interest in party platforms or ideologies, and here also he resembles his creator. When I speak of the political implications of Hawthorne's work, I do not mean that he had anything like Fenimore Cooper's concern with the operation of the American political system as defined by the Constitution, or with the policies of the national government. The political significance of Hawthorne's fiction lies rather in its bearing on the legitimacy of the entire fabric of American institutions.

Because of Hawthorne's intense preoccupation with his native region, his political thought in this broad sense is expressed in his depiction of New England society, past and present. We must keep in mind, therefore, the general tendency of the region's intellectual history. Perry Miller describes it as a process in which the overt authoritarianism of the Puritan theocracy of the seventeenth century became during the eighteenth century an urbane rationalism insepa- rable from the developing capitalism of Boston and other seaports. By Hawthorne's day this rationalism had received explicit formulation in the Unitarian doctrine that Miller calls "the ideology of . . . respectable, prosperous, middle-class Boston and Cambridge. . . ." He points out that "there was a connection between the Unitarian insistence that matter is substance and not shadow, that men are self- determining agents and not passive recipients of Infinite Power, and the practical interests of the society in which Unitarianism flour- ished." Thus although New England Unitarianism has to be de- scribed as "liberal" in its theology, "it was generally conservative in its social thinking and in its metaphysics."[2]

During the 1830s the intellectual and literary community of Boston and Cambridge was disturbed by a revolt of young intel- lectuals against this local version of American philistinism.[3] Numer- ous crusades were launched, demanding reforms not only in theology but also in the distribution of wealth, in the position of women, in the treatment of the insane, in diet, in dress, in a dozen other real or supposed abuses. Although the reformers differed widely among themselves in religious views, one group of religious radicals figures conspicuously in the history of American literature because it in- cluded a number of gifted writers. In the 1840s several members of this group—called Transcendentalists—launched a decorous experi- ment in communal living at Brook Farm in West Roxbury, near Boston. Hawthorne was close enough to this experiment to spend several months at Brook Farm (motivated primarily, to be sure, by the hope of finding a means of livelihood that would allow him to marry). Reviewers of his books would insist later on associating him with "the Concord sect" and "the Roxbury phalanx"—and Poe, in general an admirer of Hawthorne's work, would regret what he called the element of "metaphor run-mad" that Hawthorne had caught by contagion from "the phalanx and phalanstery atmos- phere."[4] Hawthorne, for his part, has Coverdale, the narrator of *The Blithedale Romance* whom some readers took to be the author's spokesman, refer with a certain irony to "the conservatives, the writers of the North American Review, the merchants, the politicians,

the Cambridge men, and all those respectable old blockheads. . . ."
But the tone of the allusion is softened by Coverdale's additional
remark that these blockheads "still . . . kept a death-grip on one
or two ideas which had not come into vogue since yesterday-
morning."[5] Hawthorne's attitude toward Brook Farm and its reformers was far from simple. If some conservative reviewers of *The
Blithedale Romance* perceived him as a radical, others welcomed the
book as a satire on foolish radicalism. According to Bertha Faust,
Hawthorne's sketch "The Celestial Railroad," reprinted in pamphlet
form, "continued for a long time to be a popular item among the
offerings of the American Sunday School Union," because it was
taken to be an attack on the current liberal trend in theology (as
indeed it was, in its genial way). The lines were hard to draw.
Emerson, leader of the theological liberals in the eyes of the orthodox,
departed from his usual coolness toward Hawthorne's work by
writing to Thoreau that "The Celestial Railroad" had "a serene
strength which we can not afford not to praise, in this low life."[6]

Serene or not, however, during the early and middle periods of
Hawthorne's writing career, the periods when he produced his best
work, he was in general highly critical of the raucous materialism and
utilitarian coarseness of American society. Lawrence S. Hall, author
of the standard monograph *Hawthorne: Critic of Society*, says that
the young writer subjected the manners and institutions around him
to "a running fire of scathing, satirical criticism," to such an extent
that he could be called "maladjusted," "censorious," even "impu-
dent."[7] One recalls almost at random the disdain for the muscular
blacksmith Robert Danforth expressed in "The Artist of the Beautiful" and the ferocious portrait of the financier-politician Jaffrey
Pyncheon in *The House of the Seven Gables*.

Judgments of this kind, however, even when they are dramatized
as conflicts between characters in fiction, are still not the most
powerful expressions of Hawthorne's thought about society and its
institutions. In order to understand his full meaning we must
recognize the challenge to all institutions that is implicit in the basic
concept of "romance" that he worked out during his long, solitary
apprenticeship. His subordination of the outer world of institutions
and observed behavior to the inner universe of private experience
places him squarely in opposition to the principles that had con-

trolled American criticism during the preceding generation. William Charvat observed that in these early decades literary judgments could not really be distinguished from social or political judgments because the leading critics acted as spokesmen for "a practically homogeneous upper class which felt itself competent to legislate, culturally, for other classes." "Religion and law together," asserted Charvat, ". . . served to create a social pattern of thought in criticism which brooked no assaults upon the political and moral order of the day."[8]

There was no question, of course, of an overt assault by poets or novelists on the principles sustaining the body politic. But the critics "thought of the fate of civil and religious institutions as interdependent." Furthermore, since "optimism was felt to be a social necessity," "pessimistic views of life" were held to be in themselves "dangerous to the social order."[9] In New England as elsewhere existing government and codes of law were believed to be legitimized by divine revelation through the Scriptures, which in turn were validated by historical evidence. An anonymous reviewer of Andrews Norton's treatise *The Evidences of the Genuineness of the Gospels* (probably Francis Bowen) explained in the *North American Review* in 1837 that "any point of doubt or difficulty" in such matters was to be handled as in a court of law:

> We found our decision either simply on the assertion of competent witnesses, or else on the tacit documents presented by well-ascertained phenomena, which stand in the relation of cause, effect, or necessary concomitant, with one or the other side of the question at issue.[10]

Once the absolute authenticity of the Scriptures was established, the rest was easy, because conservative opinion held that divine revelation guaranteed the right of private property and the existing distribution of wealth and power.[11]

For Hawthorne's New England, however, especially for the vaguely Unitarian and upper-class segment of it in which he grew up, the most urgent considerations were no longer theological in the traditional sense. The spirit of rationality had been extended to embrace all possibilities of experience through merger of Christian doctrines with Scottish common-sense realism. The implications of this official philosophy for literature are set forth with great clarity in a review of Emerson's first volume of poems by Francis Bowen, editor of the *North American Review,* subsequently professor of moral philosophy at Harvard from 1853 to 1889, and one of the best minds among "the Cambridge men."

> The publication of a volume of such poetry at the present day [wrote Bowen in 1847] is a strange phenomenon; but a stranger, still, is the eagerness with which it is received by quite a large circle of neophytes, who look down with pitying contempt on all those who cannot share their admiration of its contents. . . . How far the taste may be perverted by fashion, prejudice, or the influences of a *clique* or school, it is impossible to say; but there must be limits to all corruptions of it which come short of insanity.

The professor then clears his throat, so to speak, and begins a pontifical exposition of common-sense epistemology that ends up in a crudely exaggerated version, almost a caricature, of the neo-classical esthetic doctrine current in Britain one or two generations earlier:

> It is possible to profess admiration which one does not feel; or for the faculties to be so impaired by disease as to become insensible to their appropriate gratifications. The ear may lose its perception of the finest harmonies, the olfactory nerve may no longer be gratified by the most delicious perfumes; these would be mere defects, a loss of the sources of great enjoyment. But we cannot conceive of enjoyments being created of an opposite character. The ear cannot be trained to receive pleasure from discords, nor the sense of smell to enjoy a stench. As with the pleasures of sense, so is it with intellectual gratifications. We may never have acquired a relish for them, or we may lose it by neglect. But one cannot change the nature of things, and derive positive pleasure from that which is distasteful and odious by its original constitution.[12]

Although Bowen is ostensibly talking about diction and imagery, his real topics are epistemology and metaphysics. He is insisting that there is a material universe knowable through the senses. This universe is made up of objects having not only unchanging primary and secondary qualities, but in addition a fixed esthetic and moral valence. No unconventional associations are to be tolerated in literature because these represent an actual or implied rebellion against the settled views of the community (and therefore the human race). Furthermore, the extra-mental universe of solid and stable objects is constructed and maintained according to natural laws that are fully comprehensible by the human mind, as are indeed the mental processes of human beings themselves. The purpose of art is to represent portions of this natural universe that are pleasing to spectator or reader, in such a way as to convey moral lessons.[13]

The mode of fiction that Hawthorne in effect invented was derived from exactly antithetical premises: a different psychology, a different epistemology, a different ontology. The truth to be communicated by literature was not for him a truth about the outer universe, either physical or social, but what he called "the truth of the human heart."[14] Although Emerson seems not to have recognized it, Hawthorne's goal in fiction corresponded precisely to Emerson's description of the symbolic process: "That which was unconscious truth, becomes, when interpreted and defined in an object, a part of the domain of knowledge,—a new weapon in the magazine of power."[15] This conception transformed the reading of fiction from a frivolous pastime into a profoundly disturbing experience. What the novelist called "burrowing . . . into the depths of our common nature, for the purposes of psychological romance"[16] challenged the right to an easy conscience that was claimed implicitly for every American by the Adamic myth of rebirth and innocence in the New World. For the deep truth of the human heart that Hawthorne discovered was not perfectibility but guilt. This discovery, in turn, disturbed the cheerful surface of prosperity and contentment that was supposed to prevail in American society.[17] Reviewing *The Scarlet Letter* in 1850, Charles Hale, a recent graduate of Harvard and editor of a short-lived literary magazine called *To-Day*, acknowledged that Hawthorne's name was probably destined to immortality, but added that his work made life seem hollow and meaningless. This novel, he declared, is "an awful probing into the most forbidden regions of human consciousness. . . . " In sum, "It is gloomy from beginning to end."[18] Other reviewers professed to discover "pollution" and "mentally diseased" notions in both *The Scarlet Letter* and *The Blithedale Romance*.[19]

Anne W. Abbot, reviewing *The Scarlet Letter* in the *North American Review*, is also troubled by Hawthorne's deliberate unsettling of normal expectations. When the reader closes the book, she says, he feels giddy, as if he were "just awaking from his first experiment of the effects of sulphuric ether. The soul has been floating or flying between earth and heaven, with dim ideas of pain and pleasure strangely mingled, and all things earthly swimming dizzily and dreamily, yet most beautiful, before the half shut eye." But the beauty of these visions is dangerous: the writer's "imagination has sometimes taken him fairly off his feet, insomuch that he seems almost to doubt if there be any firm ground at all,—if we may judge from such mist-born ideas as the following." And she quotes Hawthorne's penetrating inquiry in the last chapter whether, "Philosophically considered," "hatred and love be not the same thing at bottom." These opinions, appearing in the principal critical organ of Boston and Harvard literary circles, indicate that the patrician rationalism of earlier decades could stiffen into middlebrow banalities. Mrs. Abbott recognizes Hawthorne's "wizard power over language," but on balance she condemns his work. In making these judgments she exhibits the resistance to bringing "latent or hidden motives into the light of consciousness" that according to John G. Cawelti characterizes popular culture.[20] For the representation of genuine psychological conflicts endangers the conventional view readers have of themselves and their associates.

The hostile critics were unable to identify the exact source of the uneasiness that Hawthorne's psychological romance aroused in them. They realized his immense talent, and could not really fault him for the manifest doctrine of his work. But the subtler conservatives perceived nevertheless that his fiction was implicitly subversive of the established value system. It had a decided political import in its attack both on the decadent patrician code inherited from the Federalist era (satirized, for example, in Hepzibah Pyncheon's pathetic pride in her ancestry) and on the ruthless "Go-Ahead" commercial ideology of the era of Manifest Destiny (represented by the corrupt Judge Jaffrey Pyncheon). Let us look at a few passages in which the challenge finds expression.

As early as 1832, when Hawthorne was only a few years out of college, he can be seen trying to create what Charles Feidelson, Jr., calls a "psycho-physical" universe.[21] The tale "Roger Malvin's Burial" is based on an incident in early New England history that Hawthorne says is "naturally susceptible of the moonlight of romance." Two white survivors of a fight with the Indians, both grievously wounded, are retreating through the forest. Roger Malvin, the elder, realizes that his strength is giving out and persuades his companion, Reuben Bourne, to continue without him, asking only that Bourne return later to bury his body. Reuben vows to perform this act. He sets out alone and is eventually picked up by a rescue party. When he recovers from his wounds he marries Dorcas, Malvin's daughter, but lacks the courage to tell her that he left her father to die alone and to lie unburied. This guilty secret transforms Reuben into a moody, unhappy man, neglectful or unfortunate in the cultivation of a formerly prosperous farm. Many years of struggle reduce him to bankruptcy, and the family (now including an only son, Cyrus) are forced to move westward into the wilderness for a fresh start. After they set out Reuben is drawn off his intended route by a "strange influence": he seems to hear a "supernatural voice" that calls him onward and forbids his retreat. Without quite realizing it, he leads his family to the very spot where he had left the dying Roger Malvin. Here father and son separate to hunt game for food; and Reuben, firing at some object that moves "behind a thick veil of under-growth," kills his son.[22]

The body lies beneath an oak tree to which, eighteen years before, Reuben had tied his handkerchief as a means of identifying the site. The bough to which this "little banner" had been attached is now withered (p. 403). When Dorcas, discovering her dead son, falls insensible to the ground, the withered bough

> loosened itself in the stilly air, and fell in soft, light fragments upon the rock, upon the leaves, upon Reuben, upon his wife and child, and upon Roger Malvin's bones. Then Reuben's heart was stricken, and the tears gushed out like water from a rock. The vow that the wounded youth had made, the blighted man had come to redeem. His sin was expiated,—the curse was gone from him; and in the hour when he had shed blood dearer to him than his own, a prayer, the first for years, went up to Heaven from the lips of Reuben Bourne. [p. 406]

From a common-sense standpoint this story is absurd. It represents a fictive world in which ordinary notions of cause and effect or guilt and innocence do not apply. Hawthorne's sketch offers the narrative equivalent of the irrationalities of imagery that Francis Bowen objected to in Emerson's poetry. Unless we adopt Bowen's perspective and declare the story to be simply insane, we have to find some explanation of why Reuben kills his son, to whom he is devoted. And how does the act expiate his sin? Frederick C. Crews argues persuasively that Reuben has projected his guilt upon Cyrus, who thus becomes the tainted part of Reuben's own self. When Reuben kills his son, in the non-rational domain of unconscious motivation he has cast off his burden of guilt.[23] Whether this psychological interpretation is correct or not, the passage clearly conforms to Emerson's notion of symbolic perception—which would not be made public for several years. When the elder Bourne first sees the withered bough, he wonders, "Whose guilt ha[s] blasted it?" (p. 403). A buried truth begins to rise toward consciousness: he realizes that someone's guilt is being proclaimed, although he cannot yet bring himself to acknowledge the guilt as his own. The death of Cyrus, which seems to have some connection in Reuben's mind with his failure to bury Roger Malvin's body, jolts him into a confession of his wrongdoing. At last he is able to say to his wife Dorcas, "This broad rock is the gravestone of your near kindred . . . Your tears will fall at once over your father and your son" (p. 406).[24]

But although the process by which Reuben Bourne acquires knowledge corresponds to an Emersonian theory of symbolism, what he learns about himself contradicts Emerson's belief that at its deepest level every individual self merges into the one divine Self permeating the universe, so that the basic truth about man is his divinity. The withered bough reveals to Reuben his own guilt. Yet the revelation is not, perhaps cannot ever be, complete: the burden of the guilt far exceeds his recognition of it. Indeed, Reuben's guilt is destructive in proportion as it remains closed off from his own awareness. The offense of failing to confess to his wife that he deserted her father seems minor in comparison with a darker, nameless crime that has something to do with his failure to fulfill his vow to bury Malvin's body but is never fully explained. Crews makes a strong case for regarding this unspecified offense as a repressed impulse to kill Malvin; at any rate it is excluded from Reuben's consciousness. A reader committed to the ideal of patriarchal authority might well be disturbed by this tale without quite knowing why. Reuben Bourne's impulse to kill his symbolic father is subjected to no more criticism than is his killing of his actual son.

"Roger Malvin's Burial" challenges the notion of a rational nexus of cause and effect in the outer world as well as in the inner. The force that causes the bough to fall originates outside Reuben Bourne's consciousness, yet it seems to have a significant relation to a non-empirical category—that of guilt. While avoiding any explicit reference to the doctrine of special providences, Hawthorne describes an event that resembles the numerous accounts of such providences in Puritan chronicles.

Such ambiguity proved to be important in his later work. The plot of *The Scarlet Letter* depends heavily on ontological as well as epistemological uncertainties. Many passages in the book call into question not only the public morality of an authoritarian and repressive Puritan society but the very idea of a solid, orderly universe existing independently of consciousness. For example, while Arthur Dimmesdale stands alone on the scaffold by night, his mind makes "an involuntary effort to relieve itself by a kind of lurid playfulness" (p. 151). The passage provides an unexpected touch of comic relief, and at the same time imports into the narrative a series of events having an equivocal relation to reality. Dimmesdale imagines what would happen if he should still be standing there at daybreak:

> The neighbourhood would begin to rouse itself. . . . A dusky tumult would flap its wings from one house to another. . . . Old Governor Bellingham would come grimly forth, with his King James's ruff fastened askew; and Mistress Hibbins, with some twigs of the forest clinging to her skirts, and looking sourer than ever, as having hardly got a wink of sleep after her night ride . . . Hither, likewise, would come the elders and deacons of Mr. Dimmesdale's church, and the young virgins who so idolized their minister, and had made a shrine for him in their white bosoms; which now, by the by, in their hurry and confusion, they would scantly have given themselves time to cover with their kerchiefs. [pp. 151–152]

There can be no doubt that in this passage Hawthorne is again following the procedure described by Emerson: he is using "natural facts," or rather the verbal representation of natural facts, to express truths that have been brought up into consciousness from a buried level of Dimmesdale's (and the reader's) mind. The mockery of such a figure of authority as old Governor Bellingham has been prepared for a few lines earlier by a description of Dimmesdale's hyperbolic "conceit" that the light from the lamp carried down the midnight street by "good Father Wilson," carefully labeled Dimmesdale's "professional father," represents "the distant shine of the celestial city" that Wilson might have "caught upon himself . . . while looking thitherward to see the triumphant pilgrim [that is, the spirit of Governor Winthrop] pass within its gates" (p. 150). Dimmesdale's fantasy causes him to smile and almost to laugh—and then to wonder whether he is going mad. These psychological attacks on father figures are accompanied by the sexual innuendo about the self-exposure of the young virgins.

Dimmesdale's laughter is especially significant because it suggests a connection between comedy and the release of repressed libido that would not be explicitly recognized until Freud's *Wit and Its Relation to the Unconscious*. Hawthorne is in fact depicting a mode of transcendence downward that is vaguely adumbrated in Emerson's celebrated declaration, "I embrace the common, I explore and sit at the feet of the familiar, the low."[25] But the full implications of the doctrine would become apparent only when Whitman and Mark Twain created new literary languages based on the vernacular.

More relevant to Hawthorne's procedure is the question of how much reality we should ascribe to Dimmesdale's grotesque fantasy. Does it have any substance except as a revelation of his character? From a common-sense dualistic position the answer would have to be, No. And the question arises in other forms. As every reader of *The Scarlet Letter* remembers, the scene of Dimmesdale's vigil on the scaffold culminates in his vision of "an immense letter,—the letter A,—marked out in lines of dull red light." But Hawthorne insists that this was a private vision—due "solely to the disease in his own eye and heart"—and adds, "another's guilt might have seen another symbol" in that light (p. 155). Hester Prynne, standing beside Dimmesdale, apparently does not see the "A" in the sky. Yet the author associates her with the minister in another visual experience that takes place concurrently with his seeing the "A." The "meteoric light" reveals things as they might appear on the day of judgment—that is, as revealed by the light of absolute truth. It enables both Dimmesdale and Hester to see the figure of Chillingworth at the foot of the scaffold, resembling "the arch-fiend, standing there, with a smile and a scowl, to claim his own" (p. 156). And the sexton reveals that even Dimmesdale's vision of the "A" was not entirely private by informing him next day that an unspecified number of townspeople also saw an "A" in the sky and interpreted it as standing for "Angel" in recognition of the death of Governor Winthrop.

I shall not propose a new reading of this passage, which is of course one of the two or three most celebrated in all American literature, but I do wish to call attention to the fact that it belongs to a group of scenes in which Hawthorne casts doubt almost in so many words on the common-sense notion of reality. By adopting the point of view now of one, now of another character he achieves an effect of what Leo Spitzer has called perspectivism. The implication is that the human mind can never discover the absolute truth about any extra-mental reality, but only an infinite series of partial views.[26]

For a moment Hawthorne even presents the fictive world from the point of view of Chillingworth, in his famous cry, "Let the black flower blossom as it may!" (p. 174). This climax comes at the end of the chapter called "Hester and the Physician." In the next chapter Hawthorne once again adopts Hester's point of view. She has been talking with Chillingworth by the seaside, and has retracted her promise to conceal his identity from Dimmesdale. As he walks away, she stands gazing after him,

> looking with a half-fantastic curiosity to see whether the tender grass of early spring would not be blighted beneath him, and show the wavering track of his footsteps, sere and brown, across its cheerful verdure. She wondered what sort of herbs they were, which the old man was so sedulous to gather. Would not the earth, quickened to an evil purpose by the sympathy of his eye, greet him with poisonous shrubs, of species hitherto unknown, that would start up under his fingers? Or might it suffice him, that every wholesome growth should be converted into something deleterious and malignant at his touch? . . .
>
> "Be it sin or no," said Hester Prynne bitterly, as she still gazed after him, "I hate the man!" [pp. 175–176]

It is instructive to compare this scene with that of the falling bough in "Roger Malvin's Burial." Both exemplify the process of symbolic perception, but whereas in the earlier tale the process is taking place in the mind of the implied author, here the fantasies are clearly Hester's. Hawthorne has learned to remove himself much more completely from the fictive universe, and this universe is brought correspondingly closer to absolute moral neutrality. Furthermore, Hester's act of symbolic perception is now made to serve a significant function in the plot (and thus, incidentally, to aid in the construction of a narrative pattern not limited to the length of a sketch). The natural facts that she has imagined (for example, the withered grass in Chillingworth's footsteps) become a means of definition and expression for spiritual facts—Chillingworth's evil and her recognition of it—that have not been so fully brought up into her consciousness before. As a result, she gains in knowledge both of her own heart and of Chillingworth's. The knowledge gives her energy to act: she reveals Chillingworth's identity to Dimmesdale and plans their flight to Europe. This decision in turn leads to the outburst of erotic energy during her interview with Dimmesdale in the forest. Thus, the passage I have quoted illustrates Emerson's principle that "the use of the outer creation, [is] to give us language for the beings and changes of the inward creation."[27] Let me point out once again, however, that Emerson could not recognize this aspect of Hawthorne's prose. He calls attention to exactly the same procedure in Shakespeare, but perhaps the inveterate Puritan hostility to the novel as a genre led him to consider symbolism out of the question in prose fiction.

Hawthorne makes another effective use of the transcendental distinction between mere perception of an event or object and insight into the spiritual truth behind it, in the account of Dimmesdale's delivery of his election sermon. It will be recalled that the minister had composed the sermon during the night following his return from his fateful interview with Hester in the forest: he had expended in this fashion most of the energy set free by that encounter. Hawthorne implies that the eloquence of the discourse is of a piece with Dimmesdale's self-deception—a self-deception amounting to hubris: ". . . he wrote with such an impulsive flow of thought and emotion," we are told, "that he fancied himself inspired; and only wondered that

Heaven should see fit to transmit the grand and solemn music of its oracles through so foul an organ-pipe as he" (p. 225). The organ-pipe image shows that Dimmesdale is being candid with himself about his guilt, but he still intends to maintain his hypocrisy; and one of the most extended authorial comments on the minister has already cited his determination to deliver the sermon as the most compelling evidence of "a subtle disease, that had long since begun to eat into the real substance of his character" (p. 216).

It is not clear what Dimmesdale's state of mind is when he enters the pulpit on election day, for at some unspecified time since he wrote the sermon he has decided that after delivering it he will mount the scaffold and confess his guilt before the assembled populace. There is an abundance of material here that Hawthorne might have developed if he had been disposed to make Dimmesdale the focus of the narrative. But he chooses to stay outside the minister's mind—partly no doubt for the sake of suspense; but mainly, I think, because he wishes to keep Hester at the center of his structure in this final sequence. At the same time, the decision to describe the scene from her point of view enables him to avoid any awkward questions about the sermon. He tells the reader virtually nothing about its content. We learn only that Dimmesdale delivers a message agreeable to his hearers: he foretells "a high and glorious destiny for the newly gathered people of the Lord" (p. 249). A recent critic has called the sermon, in my opinion accurately, "a ranting political oration, a hymn to American progress."[28] Hawthorne has not forgotten his consistent scorn for the nineteenth-century rhetoric of Manifest Destiny that Dimmesdale's sermon foreshadows.[29]

Hester, however, standing outside the meeting-house, cannot distinguish the words at all. The deepest meaning of the discourse is communicated to her by the "indistinct, but varied, murmur and flow of the minister's very peculiar voice." Dimmesdale's written words belong to the earlier occasion when he sat alone in his frenzy of composition, but the spoken utterance belongs to the immediate occasion of the delivery of the sermon. His discourse is now set to music that clashes with its surface meaning:

> Like all other music, it [Dimmesdale's voice] breathed passion and pathos, and emotions high or tender, in a tongue native to the human

heart, wherever educated. Muffled as the sound was by its passage
through the church-walls, Hester Prynne listened with such intentness,
and sympathized so intimately, that the sermon had throughout a meaning for her, entirely apart from its indistinguishable words. These, perhaps, if more distinctly heard, might have been only a grosser medium,
and have clogged the spiritual sense.

The passage is long and I cannot quote all of it. Basically, the message
imparted to Hester without words is a "cry of pain," "The complaint
of a human heart, sorrow-laden, perchance guilty, telling its secret,
whether of guilt or sorrow, to the great heart of mankind . . ."
(p. 243).

Although the narrative method developed by Hawthorne in the
latter part of *The Scarlet Letter* foreshadows the withdrawal of the
writer from the fictive world that we associate with Flaubert, James,
and later novelists, he was not prepared to accept the full consequences of this technical maneuver. For if the writer withdrew completely, and adopted the stance advocated by James Joyce of an
indifferent observer paring his fingernails,[30] the consequences
reached far indeed—even, in fact, to the acceptance of a fictive world
without God, that is to say without an intelligible order, an absurd
universe like that of the existentialists.[31] Hawthorne did not intend
anything so extreme. The wordless language of the heart in which
Dimmesdale communicates with Hester Prynne belongs to an order
beyond that of the senses, but far from being absurd, it transcends the
plane of the Understanding precisely because it is completely moral,
because it embodies to the fullest extent a scheme of values that
Hawthorne believed every human being must recognize intuitively as
true.[32]

When Hester Prynne by the seashore imagines Chillingworth's footsteps withering the grass, Hawthorne for a moment verges on a
complete perspectivism. Since everything outside Hester's consciousness is temporarily excluded from the fictive universe, there is no
built-in point of vantage from which she can be judged. The reader is
given no indication, for example, whether the author approves or
disapproves of her expression of hatred for Chillingworth. But
Hawthorne's more usual method in *The Scarlet Letter* is illustrated

by the scene of Dimmesdale's election sermon. Here the narrative voice speaks for a consciousness that is aware of Hester's thoughts and feelings, and at the same time is given authority to deliver judgments such as the suggestion that mere words might have been a "grosser medium" capable of clogging her "spiritual sense" (p. 243). In this fashion a noumenon is established that contains the transcendent meanings constituting Hawthorne's realm of pyschological romance.[33]

The persistence of this cluster of meanings that collectively make up the truth of the heart is the feature that gives coherence to Hawthorne's career as a novelist. Although it is not wholly unrelated to his experiments with point of view, it remains constant despite his shifts in technique. From the ventures in perspectivism in *The Scarlet Letter*, for example, Hawthorne turns only a year later, in *The House of the Seven Gables*, to the use of a narrative voice that moralizes freely and indeed dominates the fictive world. Then in *The Blithedale Romance* he makes his only attempt to present a book-length story through the eyes of a first-person narrator—with the result, I might add, that critics are still arguing about whether or not the reader is supposed to accept this narrator, the dilettantish poet and man-about-town Miles Coverdale, as a fully reliable reporter of events in the story.[34] Finally, in *The Marble Faun*, Hawthorne's last published novel, he returns to a loose third-person narrative method that allows him to enter characters' minds when he wishes, but to stay out also for such purposes as leaving intact the mystery surrounding Miriam's past or Hilda's whereabouts during her temporary disappearance.

These shifts in technique are usually directed toward blurring the line between actual and imaginary, outer and inner realms. In *The House of the Seven Gables*, for example, during one entire chapter (ironically entitled "Governor Pyncheon") the only character present is the corpse of Judge Jaffrey Pyncheon, whose death, possibly caused by the operation of a curse placed upon his ancestor two centuries earlier, has prevented his nomination for the governorship. The action of this chapter, if action it can be called, is a kind of rhetorical dance of triumph before his dead body by the disembodied narrative voice. Yet even here Hawthorne's preoccupation with the problem of reality forces itself oddly into the discourse. As twilight darkens the room where Judge Pyncheon's corpse sits, the narrative voice approaches hysteria: "An infinite, inscrutable blackness has annihilated sight! Where is our universe? All crumbled away from us; and we, adrift in chaos, may hearken to the gusts of homeless wind, that go sighing and murmuring about, in quest of what was once a world!" (pp. 276–277).

The insubstantial realm of the imagination thus contains dangers as well as consummations. The only guarantee of reality is the truth of the heart. Before Hawthorne's marriage to Sophia Peabody, in 1840, he had written to her:

> Indeed, we are but shadows—we are not endowed with real life, and all that seems most real about us is but the thinnest substance of a dream—till the heart is touched. That touch creates us—then we begin to be—thereby we are beings of reality, and inheritors of eternity.[35]

In *The House of the Seven Gables,* it is Phoebe Pyncheon who offers this kind of assured contact with reality. The narrative voice declares:

> She was real! Holding her hand, you felt something; a tender something; a substance, and a warm one; and so long as you should feel its grasp, soft as it was, you might be certain that your place was good in the whole sympathetic chain of human nature. The world was no longer a delusion. [p. 141]

Hawthorne uses the reality-principle in Phoebe to provide the resolution of the ancient conflict between Pyncheons and Maules, and thus the denouement of his plot. At the climax of the action, the ordinary assumptions about what is actual and what is imaginary are obliterated when the declaration of love between the heroine Phoebe Pyncheon and young Holgrave the daguerreotypist transforms the fictive universe by setting in motion a renewal of the life force:

> . . . it was in this hour [declares the narrative voice], so full of doubt and awe, that the one miracle was wrought, without which every human existence is a blank. The bliss, which makes all things true, beautiful, and holy, shone around this youth and maiden. They were conscious of nothing sad nor old. They transfigured the earth, and made it Eden again, and themselves the two first dwellers in it. The dead man, so close

beside them, was forgotten. At such a crisis, there is no Death; for
Immortality is revealed anew, and embraces everything in its hallowed
atmosphere. [p. 307]

This is perhaps a little too ecstatic in the manner of sentimental
fiction. But Hawthorne redeems the stereotyped material by lifting it
above mere boy-meets-girl triviality to the level of a cosmic process:
first by the Virgilian references to a yellowed bough on the Pyncheon
Elm in front of the house, "like the golden branch, that gained Aeneas
and the Sibyl admittance to Hades" and guaranteed their safe return
(p. 285), and then by breaking down all distinction between inner and
outer, so that flowers suddenly bursting into bloom in an angle of the
roof and the lovers' whispered conversation in the dark house are
parts of a continuum that is neither physical nor mental in the
ordinary sense of these terms. They exist on the same ontological
plane, which also includes the adventures of Aeneas and the Sibyl and
the melancholy story of Alice Pyncheon, known to the reader through
Holgrave's fictional version of it. The meaning of the passage has
nothing to do with either a speculative moral philosophy or an
empirical science of mind. What makes things happen in this scene is
not a nexus of psychological cause and effect but the pattern of a
myth.

Furthermore, although Hawthorne's diction is euphemistic, he
refers explicitly to the life-force that in sentimental fiction was almost
completely buried under various conventions and decorums. Even
this truncated myth affirms unconscious energies with enough force
to overcome at least partially the trivializing effect of the happy
ending. It converts the country estate and the handsome fortune of
Judge Pyncheon inherited by the inmates of the old house into tokens
of the animal vitality and sexual drive that each generation of
mankind is forced to relinquish to its successor.

In the end, however, the importance of Hawthorne's efforts to
construct an ontologically ambiguous fictive world reached beyond
the philosophical content of his work, which was simply one aspect
(in Ernest Tuveson's phrase) of the general Romantic discovery of the
imagination as a means of grace.[36] The problem was how to transcend
raw, unmediated observation and experience. Stripped of meta-

physical implications by a process analogous to the secularization of
religious doctrines, Hawthorne's innovations in narrative technique
became the point of departure for the career of Henry James.

The Madness of Ahab

The axis of meaning in *Moby-Dick* is the contrast between what the White Whale means to Ahab and what it means to Ishmael. Indeed, in probing into this topic Melville discovers that there are as many interpretations of that Leviathan as there are consciousnesses to interpret it. As a result, the book is highly inconsistent, or perhaps one should say unconventional, in its management of narrative point of view. For example, Melville does not hesitate to include scenes composed entirely of stage directions and dialogue in a story that is ostensibly being told in the first person by one of the characters. And Chapter 99 is a kind of perspectivist ballet in which a procession of observers read meanings into the gold doubloon Ahab has nailed to the mast that are simply projections of their own buried selves. Since all the manipulations of point of view come in sections composed after Melville began rewriting the book under the influence of his discovery of Hawthorne's *Mosses from an Old Manse*, one is inclined to believe that the perspectivism in *Moby-Dick* is part of the debt Melville declared he owed to Hawthorne. But even this major item of technique is negligible in comparison with the "power of blackness" that Melville perceived in Hawthorne, the "grand truth" of "saying NO! in thunder." He found the same truth in Shakespeare, who, he asserted, "Through the mouths of the dark characters of Hamlet, Timon, Lear, and Iago . . . craftily says, or sometimes insinuates, . . . things which we feel to be so terrifically true that it were all but madness for any good man, in his own proper character, to utter, or even hint of them. Tormented into desperation, Lear, the frantic king, tears off the mask, and speaks the sane madness of vital truth."[1]

In order to explain what that dark truth was, Melville needed a great deal of sea room, no less in fact than the whole of his immense book. But along the way we are given various hints which amount to the declaration that the universe is not controlled by an all-powerful and loving God, but by an omnipotent power resembling the Devil of Calvinist theology. Ishmael (who often acts as spokesman for the author) is at first strongly attracted to this view, but in the end rejects it, and Melville also backs away from it by declaring that Ahab is mad.

The ravings of a madman are not to be taken as a guide to life. Or are they? For this madman is a special case, a heroic figure, "a grand, ungodly, godlike man" (p. 76). The author suggests more than once that Ahab is uttering a higher truth transcending mere everyday common sense. The question at issue is, whether a man who respects himself and is not a mere coward can honorably submit to the mundane principles that underlie society and its institutions. The world in question is Melville's own. And although his debate with himself is scored for full orchestra and a chorus of many voices, as contrasted with Hawthorne's quiet chamber music, the reader would not need Melville's explicit statement to recognize that he is handling the same theme. The question demands extended consideration before it can be answered. I propose to come at it by trying to determine what Ahab's madness means: that is, in what ways Melville elaborates or qualifies the basic proposition that the controlling power in the universe is evil, by declaring Ahab, who holds a view something like this, to be insane. The inquiry is particularly difficult because the book surrounds Ahab with a cloud of ambiguity. For the fact that he is mad does not at all mean that what he says is mere gibberish. Like Lear, he is suffering from a "sane madness," and is a heroic figure. Ahab often seems inscrutable in his proud isolation, yet he unpacks his heart in elaborate soliloquies; and on occasion the first-person narrator Ishmael enters Ahab's mind to conduct detailed analyses of his thoughts and feelings. In these ways Melville offers several accounts of the origin and nature of his protagonist's madness that to some extent overlap and contradict one another. In other words the insanity is overdetermined. I take this to mean that Melville is trying to express in Ahab an impulse originating below the threshold of his own consciousness. I have no intention of trying to psychoanalyze Melville in order to determine just what repressed conflicts were rising to the surface in this fashion, but shall confine myself to considering how Ahab's madness is dealt with in the novel. In order to prevent misunderstanding, let me add that my inquiry does not imply a denial of Melville's magnificent accomplishment in his masterpiece. What I am saying about *Moby-Dick* is essentially what T. S. Eliot says about *Hamlet:* it is "full of some stuff that the writer could not drag to light, contemplate, or manipulate into art." Melville has not found an "objective correlative" fully adequate to the emotion he was trying to express. This emotion "is inexpressible," and therefore remains "in *excess* of the facts as they appear" in the novel. In order to understand Ahab's state of mind fully, we "should have to understand things which [the author] did not understand himself."[2]

Several lines of development that can be traced in Melville's thought converge in the depiction of Ahab's madness. First, there are intimations in the writer's earlier work, at least from *Mardi* (1849) onward, that insanity can be a means of access to higher truth. Babbalanja, a character in *Mardi*, introduced as a philosopher, is sometimes possessed by a devil named Azzageddi who violently denounces social injustices in various domains within the archipelago of Mardi—especially the luxurious self-indulgence of a certain king whose subjects are condemned to live in rags and misery.[3] Later one of the other characters calls Babbalanja's ranting "madness" (p. 616). In *Redburn* (also published in 1849), a seaman named Jackson dominates the crew of a merchant vessel in the north Atlantic by means of a peculiar personal force of character and will, although he "seemed to be full of hatred and gall against every thing and every body in the world; as if all the world was one person, and had done him some dreadful harm, that was rankling and festering in his heart."[4] This generalized hostility toward an ill-defined target, together with a mysterious power over other men, anticipates some traits of Ahab.

A quite different aspect of Melville's thought that entered into the depiction of Ahab's madness was his acquaintance with contemporary psychological theory. A revolution had occurred in this field of knowledge in Europe around the turn of the nineteenth century, and had been quickly noted in the United States. During the two years between the composition of *Redburn* and that of *Moby-Dick* Melville seems to have become aware particularly of the discussion of monomania, for he makes use of information derived from this source in depicting Ahab's madness. The term, and with it the notion of a disease with specific symptoms, had become current only recently. Isaac Ray, a leading American psychiatrist, wrote in 1838 that it had been introduced by the noted French physician J. E. D. Esquirol; and this opinion is confirmed by the British psychiatrist James C. Prichard, who noted in 1835 that the term had appeared in English "within a few years."[5] The OED places the first use of the word in 1823. It had gained currency quickly in England and America because of a controversy over the validity of the plea of "moral insanity," of which monomania was considered to be one type, as a defense in criminal prosecutions.[6] Not surprisingly, the psychological terminology available to Melville was confused. Although a literal interpretation of moral insanity (as of *"manie sans délire"*) would seem to rule out delusion, Chief Justice Lemuel Shaw of the Supreme Judicial Court of Massachusetts (whose daughter Herman Melville would marry in 1847) stated in an opinion written in 1844 that in cases of monomania, "The conduct may be in many respects regular, the mind acute, and the conduct apparently governed by rules of propriety, and at the same time there may be insane delusion, by which the mind is perverted." In such cases, "the mind broods over *one idea* and cannot be reasoned out of it."[7] This implies a localized delusion, confined to a sharply defined segment of the mental horizon, but still a derangement of cognition. Ray also seems, rather inconsistently, to reject the notion that monomania is a form of moral insanity or *manie sans délire* when he concedes that in monomania, although "the mind is not observed to have lost any of its original vigor," the patient suffers from delusion concerning one topic (p. 124).

The British common-law test of insanity had been set forth in 1843 in the widely influential McNaughton decision, which used as a criterion the individual's ability to distinguish right from wrong.[8] The notion of moral insanity upset this principle by implying that a man might be fully aware of the immorality of a possible course of action (that is, might have normal cognition), yet suffer from a disease of his "moral" faculties (emotions and will) that rendered him unable to resist his impulse to follow the admittedly wrong line of behavior. Ray, for example, quotes the German psychiatrist Johann C. Hoffbauer to the effect that "The maniac may judge correctly of his actions without being in a condition to repress his passions and to abstain from the acts of violence to which they impel him" (p. 131). Elsewhere Ray asserts that a sufferer from "partial moral mania," while "he retains the most perfect consciousness of the impropriety and even enormity of his conduct . . . deliberately and perseveringly pursues it" (p. 140). Amariah Brigham, director of the New York State Lunatic Asylum at Utica, made an interesting observation in the *American Journal of Insanity* in 1848 concerning confusion in the public mind about the novel idea of moral insanity. But Brigham's language is not really clearer than that of Justice Shaw, for his interpretation eliminates the distinction between non-cognitive and cognitive processes by lumping them together as functions of the brain:

> The disbelief in a kind of insanity that does not disturb the intellect, arises perhaps from the common phraseology, that the affections, passions, and moral qualities, have their seat in the *heart* and not in the brain [italics in the original], and therefore are not likely to be disordered by disease of the latter organ. But in fact the orderly manifestations of our moral faculties, our affections, and intellectual powers, are alike dependent on the healthy state of the brain. *The heart has nothing to do with either.* [Italics in original][9]

Ishmael, evidently serving as a mouthpiece for Melville, is explicit in his assertion that Ahab is suffering from a disease of his moral powers not his intellect:

> Now, in his heart, Ahab had some glimpse of this, namely: all my means are sane, my motive and my object mad. Yet without power to kill, or change, or shun the fact; he likewise knew that to mankind he

did long dissemble; in some sort, did still. But that thing of his dis-
sembling was only subject to his perceptibility, not to his will determinate. [p. 161]

"Perceptibility" and "will determinate" are, however, quite unclear,
especially "perceptibility," which seems to be a synonym for "cognition." In fact, current psychological theory, including the notion of
monomania, was of little use to Melville in depicting the texture of
Ahab's mental processes. What the theory did offer him was not
unlike what he derived from the various accounts of the sinking of the
whale ship *Essex* off Nantucket by a whale in 1820, and newspaper
reports of the sinking of the ship *Ann Alexander* (out of New Bedford)
by a sperm whale while *Moby-Dick* was in press: it suggested plot
material and helped make Melville's story acceptable to readers.[10] The
scientific conception of monomania indicated that as a sufferer from
this disease, Ahab could pursue an insane course of vengeance against
the White Whale (one which would be morally reprehensible if he
were sane), yet present the appearance of sanity to the owners of the
Pequod and retain to the full his capacity to command the vessel and
dominate the crew. It was also necessary that despite Ahab's madness,
he should be capable of the impressive intellectual feat of determining where the whale would most probably reappear by analyzing
ocean currents and records of previous sightings. Thus the novelist
established the empirical plausibility of a plot that sends an actually
crazy but apparently sane skipper across the oceans of the world in a
successful search for a specific giant whale.

Current psychological theory aided Melville also by maintaining
(in Ray's words) that mania might be caused by "external shocks" as
well as by "moral shocks."[11] This idea could have suggested attribut-
ing Ahab's madness to a specific injury inflicted by the White
Whale. Ishmael describes the incident, but since it occurred before he
had any contact with the Captain, he must be assumed to have learned
about it from members of the crew of the *Pequod* who had sailed
under Ahab on the previous voyage. Yet Ishmael's account is one of
the most vivid passages in a book with many vivid passages; its
sensory detail suggests first-hand experience. This may be implausible from a technical standpoint but Melville was prepared to place
even more strain on narrative perspective when the situation de-

manded it, and here he was dealing with a moment of crisis in the story. For the first meeting of Ahab with Moby Dick reveals Ahab's state of mind at a time when he had not yet succumbed to monomania. Up to that moment he had faced only the hardships and privations that were the lot of every man who rose to command in the whale fishery. There was no question as yet of the mutilation that was literally to drive him mad. Yet Ishmael depicts Ahab as being mastered by rage of heroic, even monstrous intensity. The implication must be that other veteran whale-hunters who did not feel such frustration and hostility were different from Ahab in temperament, coarser, lacking his depth. (Bluff Captain Boomer, the British skipper of the *Samuel Enderby*, seems placed in the story for the express purpose of pointing up the contrast. He and his surgeon make a joke of how when Boomer lost an arm in pursuing Moby Dick, the surgeon nursed him back to health by drinking rum punches with him until three o'clock every night for weeks (p. 367)). The fact that the fury Ahab displayed in this first encounter with the White Whale was generated by widely shared experiences gives it more general meaning than the subsequent madness.

Yet if we turn to Ahab's earlier life for a clue to his state of mind, we find only meager information. Ahab's father is never mentioned, and of his mother we know only that she was "crazy," that she gave her infant the name of a wicked Old Testament king, and that she died soon after his birth. Further, a shadowy Indian sorceress predicted he would justify the name (p. 77). He was injured in some fashion off Cape Horn; he became involved in an altercation "afore the altar in Santa," perhaps by spitting in a ritual vessel (p. 87); and "past fifty" he married a "young girl-wife, and sailed for Cape Horn the next day" (p. 443). "Not three voyages wedded," she bore him one child (p. 77). To these bare facts must be added Ahab's one moment of reminiscence to Starbuck, just before the first lowering for Moby Dick, concerning his "forty years of continual whaling"—of which he has "not spent three ashore." For much of that time, when at sea he has suffered "the walled-town of a Captain's exclusiveness," the "Guinea-coast slavery of solitary command": not to speak of such lesser hardships as constant "dry salted fare—for emblem of the dry nourishment of my soul!" More significant has been "the madness, the frenzy, the boiling blood, and the smoking brow, with which, for a thousand lowerings old Ahab has furiously, foamingly chased his prey—more a demon than a man . . . " (pp. 443–444).

In order to prepare the reader for Ahab's first meeting with the whale, Melville recounts rumors long circulated among whalemen of the creature's huge size and his "unexampled, intelligent malignity"—culminating in the widespread belief of sailors that he had supernatural powers (pp. 158–159). In this way Melville implies that Ahab's reactions were not unique; there were others who had similar emotions about the White Whale:

> Judge, then [says Ishmael], to what pitches of inflamed distracted fury the minds of his more desperate hunters were impelled, when amid the chips of chewed boats, and the sinking limbs of torn comrades, they swam out of the white curds of the whale's direful wrath into the serene, exasperating sunlight, that smiled on, as if at a birth or bridal. [p. 159]

"It is not probable that this monomania in him took its instant rise at the precise time of his bodily dismemberment" (p. 160). Rather, it was the outcome of a process: over a period of time Ahab made the White Whale the focus of exasperations that had accumulated in him as a widely if not universally shared human experience. In this interpretation (which I think Melville clearly intended) the insanity consists in a cognitive change that focuses hostility and resentment previously directed against life and the universe in general, on the single tangible and accessible adversary. Melville speaks of "deliriously transferring" (p. 160) the idea of all the evil in the universe to the whale; at least since Robert Burton's *Anatomy of Melancholy* (1621), "delirium" had referred to the cognitive errors of the insane.[12]

Evidently, however, Ahab's madness had deeper implications for Melville than any mere enumeration of symptoms can convey. Although the "final monomania" may seem superficially to bear a direct relation to Ahab's unique experience of mutilation by his monstrous foe, if we are to follow the writer's full intention we must (with some sacrifice of strict logic) allow this madness a representative significance comparable in scope to Ahab's accumulated resentments down to his first encounter with the whale. In either case Melville has in mind an aristocratic distinction between the insight of an elite of deep men and the insensitivity of average mediocrity, represented by the three mates and Captain Boomer. In this fictive

world, as Armin Staats points out, all men have suffered in Adam's
fall the loss of paradise and are condemned to earn their bread in the
sweat of their brows,[13] but it should be added that only exceptionally
endowed men have the insight to recognize the injustice of the human
condition and the courage to defy the cosmic powers that oppress
mankind.

Ahab must thus be accorded heroic status even before he challenges the whale and suffers the dismemberment that drives him
unequivocally mad. Yet his madness raises him to a still higher,
mythical level. The doctors may speak of the surprising powers of
reasoning to be observed in some monomaniacs, may even demonstrate that Hamlet and Lear are textbook cases,[14] but this clinical
approach cannot possibly do justice to the grandeur that Melville sees
in such tragic protagonists, and wishes to confer on Ahab. Using
Ishmael as a mouthpiece, he surrounds Ahab in his monomania with
an aura of associations drawn from the entire history of Western
culture:

> The White Whale swam before him as the monomaniac incarnation
> of all those malicious agencies which some deep men feel eating in them,
> till they are left living on with half a heart and half a lung. That intan-
> gible malignity which has been from the beginning; to whose dominion
> even the modern Christians ascribe one-half of the worlds; which the
> ancient Ophites of the east reverenced in their statue devil;—Ahab did
> not fall down and worship it like them; but deliriously transferring its
> idea to the abhorred white whale, he pitted himself, all mutilated,
> against it. All that most maddens and torments; all that stirs up the lees
> of things; all truth with malice in it; all that cracks the sinews and
> cakes the brain; all the subtle demonisms of life and thought; all evil,
> to crazy Ahab, were visibly personified, and made practically assailable
> in Moby Dick. He piled upon the whale's white hump the sum of all
> the general rage and hate felt by his whole race from Adam down; and
> then, as if his chest had been a mortar, he burst his hot heart's shell
> upon it. [p. 160]

To make the symbolic castration of the reaped leg completely
explicit, Ishmael adds that at some unspecified time after Ahab had
recovered sufficiently to get about with a leg of ivory, the artificial leg
became "violently displaced" by some "inexplicable, unimaginable
casualty," giving him "an agonizing wound" in the groin (p. 385).

At the surface level of plot, then, a lifetime at sea characterized by hardship, loneliness, and privation (including sexual privation) had built up in Ahab before his first encounter with Moby Dick frustration and rage, shared to some extent by other whaling captains (but not by all). The injury he receives from the White Whale brings his resentments to a head and over a period of some months engenders in him a full monomania. In depicting this second stage, the onset of a mental disease with a medical name and a checklist of symptoms, Melville resorts to elaborate clusters of imagery, of which I wish to examine two particularly rich and complex examples. In explaining how, "on the homeward voyage, after the encounter . . . the final monomania seized" Ahab, the author virtually abandons the conventional device of using Ishmael as a first-person narrator and takes over the narrative himself, with omniscient powers of entering the mind of the Captain. Although "his mates thanked God the direful madness was now gone; even then, Ahab, in his hidden self, raved on."

> Ahab's full lunacy subsided not, but deepeningly contracted; like the unabated Hudson, when that noble Northman flows narrowly, but unfathomably through the Highland gorge. But, as in his narrow-flowing monomania, not one jot of Ahab's broad madness had been left behind; so in that broad madness, not one jot of his great natural intellect had perished. That before living agent, now became the living instrument. If such a furious trope may stand, his special lunacy stormed his general sanity, and carried it, and turned all its concentrated cannon upon its own mad mark; so that far from having lost his strength, Ahab, to that one end, did now possess a thousand fold more potency than ever he had sanely brought to bear upon any one reasonable object. [p. 161]

It seems to me that overdetermination has now set in. "A thousand fold" is bombast. Furthermore, Melville himself realizes he is obscuring his thought in a fog of rhetoric, and apologizes for the multiplication of images and analogies. Within the space of a few lines Ahab's condition before his mutilation is called both his "broad madness" and his "general sanity." Even if we assume that Melville is using "agent" in a now obsolete sense to designate what we should call the principal who commands the agent, the reference of pronouns is blurred. And although the metaphors and analogies are in large part merely ornamental, they do have implications that overlap and thus leave an impression of contradicting each other.[15]

But these difficulties are minor in comparison with those raised by the following paragraph. Developing the implication that Ahab's lifetime of hard, dangerous labor has rendered him sexually impotent, Ishmael resorts to perhaps the most cryptic metaphor in the book. He invites his readers to descend with him into the buried baths beneath the Hôtel de Cluny in Paris. Here, "far beneath the fantastic towers of man's upper earth," Ahab's

> root of grandeur, his whole awful essence sits in bearded state; an antique buried beneath antiquities, and throned on torsoes! So with a broken throne, the great gods mock that captive king; so like a Caryatid, he patient sits, upholding on his frozen brow the piled entablatures of ages. Wind ye down there, ye prouder, sadder souls! question that proud, sad king! A family likeness! aye, he did beget ye, ye young exiled royalties; and from your grim sire only will the old State-secret come. [p. 161]

Not surprisingly, this passage has given commentators pause. Brodhead by-passes it completely. But Staats confronts it squarely and assimilates the mythological figures persuasively to Melville's critique of society. After linking the captive king with Kronos-Saturn, imprisoned in Tartarus by Zeus-Jupiter—a theme that Melville would develop fully in *Pierre*—Staats declares that "Ahab, after his castration, rebels in monomaniac despair against the new god of heavenly fire." And he continues:

> The "prouder, sadder souls" among the readers are the children of Saturn, who are condemned and banished from the serenity and self-complacent arrogance of the Olympians. . . . The Biblical Ishmael, cheated out of his paternal heritage, exiled from civilization, is the prototype of these "young exiled royalties." The state secret which [Ishmael's] melancholic imagination seeks to fathom is awareness of the downfall of an Olympian order, a repressive civilization whose supreme achievements—"however grand and wonderful" [*Moby-Dick*, p. 161]—are based on the repression and exploitation of mankind since Adam, the shackling of the elementary forces of nature and of man, in so far as through his sexuality he participates in them. [p. 113]

This seems to me not only ingenious, but in considerable part true, as a statement of intuitions that rose into Melville's consciousness while he was writing the book. But at the same time it strikes me that Staats has had to treat the passage as a do-it-yourself kit requiring the consumer-reader to contribute much labor to arranging and assembling the parts placed before him.

Looking back upon a later stage of Ahab's physical convalescence, when his monomania had become fully established, Ishmael attempts to analyze the Captain's mental processes by using a pseudo-technical apparatus of hypostatized faculties. The passage is so central to the present inquiry that I shall quote from it at length:

> Often, when forced from his hammock by exhausting and intolerably vivid dreams of the night, which, resuming his own intense thoughts through the day, carried them on amid a clashing of phrensies, and whirled them round and round in his blazing brain, till the very throbbing of his life-spot became insufferable anguish; . . . a wild cry would be heard through the ship; and with glaring eyes Ahab would burst from his state room, as though escaping from a bed that was on fire. Yet these, perhaps, instead of being the unsuppressable symptoms of some latent weakness, or fright at his own resolve, were but the plainest tokens of its intensity. For, at such times, crazy Ahab, the scheming, unappeasedly steadfast hunter of the white whale; this Ahab that had gone to his hammock, was not the agent that so caused him to burst from it in horror again. The latter was the eternal, living principle or soul in him; and in sleep, being for the time dissociated from the characterizing mind, which at other times employed it for its outer vehicle or agent, it spontaneously sought escape from the scorching contiguity of the frantic thing, of which, for the time, it was no longer an integral. But as the mind does not exist unless leagued with the soul, therefore it must have been that, in Ahab's case, yielding up all his thoughts and fancies to his one supreme purpose; that purpose, by its own sheer inveteracy of will, forced itself against gods and devils into a kind of self-assumed, independent being of its own. Nay, could grimly live and burn, while the common vitality to which it was conjoined, fled horror-stricken from the unbidden and unfathered birth. Therefore, the tormented spirit that glared out of bodily eyes, when what seemed Ahab rushed from his room, was for the time but a vacated thing, a formless somnambulistic being, a ray of living light, to be sure, but without an object to color, and therefore a blankness

in itself. God help thee, old man, thy thoughts have created a creature in thee; and he whose intense thinking thus makes him a Prometheus; a vulture feeds upon that heart for ever; that vulture the very creature he creates. [pp. 174–175]

This passage also has given difficulty to commentators. Both Brodhead and Staats avoid it. The most determined efforts to interpret it with which I am familiar are those of Paul Brodtkorb, Jr., and Robert Zoellner, neither of whom seems to me to discover a usable meaning in Melville's tormented prose. Brodtkorb observes accurately: "The passage is full of complex abstractions, qualifications, extensions, synonyms with subtle distinctions implied between them, and second thoughts. . . . there are loose ends to the precision of Ishmael's analysis; he makes too many abstract synonyms."[16] In order to account for the discrepancies between Ishmael's generalizations about Ahab and the character's actual behavior, Zoellner postulates a "psychic paradigm" in which a "pre-literary" or "pre-textual Ahab" is modified by the events of the plot.[17] But this seems a desperate measure somewhat like the "Cycle and epicycle, orb in orb" of the predecessors of Copernicus. I think the upshot of the matter is that Melville found he could not construct a satisfactory model of Ahab's mind by using an apparatus of faculties even with the addition of uncanonical faculties such as "life-spot" and "living principle" and "common vitality." His effort at a technical analysis yielded a blank analogous to the "colorless, all-color of atheism" mentioned in Ishmael's dissertation on "The Whiteness of the Whale" (Chapter 42, p. 169).

Paradoxically, this failure reveals the subtlety of Melville's insight into mental processes and his responsibility to observed psychological fact. He is superior to the professionals of his day in recognizing (even though inarticulately) that the psyche is not a structure of faculties and cannot be imagined as functioning mechanically, no matter how complex the mechanism is taken to be. For this reason, the imagery that he draws even from so humble a source as folklore to represent Ahab's madness expresses his meaning more effectively—and is actually no less accurate as a description of mental processes—than the vocabulary of contemporary science. The ellipse in the passage just quoted marks the omission of a peculiarly vivid cluster of images: the narrator declares that the "spiritual throes in him [Ahab] heaved his being up from its base, and a chasm seemed opening in him, from which forked flames and lightnings shot up, and accursed fiends beckoned him to leap down among them"—that is, into "this hell in himself" that "yawned beneath him . . . " (p. 174).[18]

The flames of hell and the accursed fiends have their source in a tradition of demonology in popular culture reaching back to medieval Christian folklore, as exemplified, for example, in the paintings of Hieronymus Bosch, that had been kept alive in gothic fiction. Closely related to such material, although not explicitly involving demonic possession, are Ahab's ranting speech when the corpusants appear at mastheads and yardarms during the storm (pp. 416–418) and his ceremony of magnetizing a fresh needle when atmospheric electricity reverses the poles of the compass (p. 425). The twentieth-century reader is likely to assume that at these moments Ahab is play-acting, like Mark Twain's Connecticut Yankee staging "miracles" to impress King Arthur and his subjects. But this assumption is called in question by the fact that Ishmael, and Melville as well, is prepared to accept without skepticism Fedallah and the Oriental boatmen whom Ahab conceals below decks (presumably to avoid arousing the resistance of the crew of the *Pequod* to his project of revenge). Ahab reveals the boat-crew only with the first lowering against the White Whale. On this occasion, says Ishmael,

> . . . what it was that inscrutable Ahab said to that tiger-yellow crew of his—these were words best omitted here; for you live under the blessed light of the evangelical land. Only the infidel sharks in the audacious seas may give ear to such words, when with tornado brow, and eyes of red murder, and foam-glued lips, Ahab leaped after his prey. [pp. 192–193]

Fedallah is in some sense supernatural, a devil haunting Ahab and exercising control over him (e.g., p. 199). Parenthetically, however, it should be noted that one of the most amusing touches of humor in this grim book is Stubb's sardonic joking to his boat-crew about the "five more hands come to help us—never mind from where—the more the merrier" (p. 188). Melville seems only half serious in his use of these gothic paraphernalia; and not even his most devoted admirers are inclined to defend them nowadays.

Still another set of images that Melville invokes to depict Ahab's madness has been treated with more respect by the commentators, no doubt because it belongs to the tradition of high culture. This is the conception traceable at least as far back as classical Greek thought that madness is supernatural but benign rather than malevolent. Throughout history the madman had been classed among the wretched of the earth. When he could not be cared for by his relatives, he had been allowed to wander about begging, or had been confined in the same prison with paupers and criminals.[19] But even in his rags and his misery, the lunatic had been viewed with a touch of superstitious awe, and this hint of mystery in the unreason of madness was caught up and developed in the Romantic notion of a small company of supremely gifted geniuses whose abilities and accomplishments could not be accounted for within the intellectual horizon of ordinary humanity.[20]

Goethe, recalling Socrates' *daimon,* had explained the superior powers of these extraordinary mortals by saying they possess a demonic element, capable of evil as well as good because it operates without regard to commonplace criteria of right and wrong, yet responsible for all the supreme achievements of the race.[21] The relation between genius and madness was elaborated further by Carlyle, whose half-grotesque hero Teufelsdröckh writes:

> Witchcraft, and all manner of Spectre-work, and Demonology, we have now named Madness and Diseases of the Nerves. Seldom reflecting that still the new question comes upon us: What is Madness, what are Nerves? Ever, as before, does Madness remain a mysterious-terrific, altogether *infernal* boiling-up of the Nether Chaotic Deep, through this fair-painted Vision of Creation, which swims thereon, which we name the Real. Was Luther's Picture of the Devil less a Reality, whether it were formed within the bodily eye, or without it? In every the wisest Soul lies a whole world of internal Madness, an authentic Demon-Empire; out of which, indeed, his world of Wisdom has been creatively built together, and now rests there, as on its dark foundations does a habitable flowery Earth-rind.[22]

Viewed from this perspective, Ahab's madness can seem a mark of distinction, an aspect of the "disease" that Ishmael declares to be inseparable from "all mortal greatness" (p. 71). The disease of greatness invariably involves suffering. When Ahab first shows himself on deck, he stands before the crew "with a crucifixion in his face; in all the nameless regal overbearing dignity of some mighty woe" (p. 111). In these paradoxes, Melville believed, lay the essence of tragedy. And he was aware that he had identified himself to a considerable degree with the tragic hubris of Ahab's defiance of the gods of this world. But although he knew (as he wrote Hawthorne) that he had written a wicked book, he also understood the meaning of catharsis: the archetypal criminal is condemned to the archetypal punishment of death for his insane presumption, and the creator of the fictive world in which justice is enforced in this fashion could feel "spotless as a lamb."[23]

The captive king in the Halls of Thermes is apparently condemned to eternal imprisonment; Melville does not imagine an unbinding of his Prometheus. Yet he does portray the escape of a younger brother and surrogate, Ishmael, with whom on the surface he is more closely identified than he is with Ahab, for Ishmael is not only first-person narrator, often indistinguishable from the author in learning and command of language, but is also identical in age and relative inexperience with the young Melville who shipped in the *Acushnet* for the actual whaling cruise that provided the basic material for the book. In short, as has been often pointed out, this story has two protagonists.[24] Much of the power of the novel is generated by the fact that Melville was of two minds while he was writing it—that he had, like Ishmael, a strong impulse to identify himself with Ahab, and only by a relatively narrow margin was capable of the act of will that ruptured this identification and reestablished his control over his materials. Even so, Ishmael, the force of health and moderation, is a shadowy figure beside Ahab.[25] The first-person narrator disappears for long sequences in the course of the book, and has little concrete evidence to offer in support of his essentially ideological declaration in "A Squeeze of the Hand":

> For now . . . by many prolonged, repeated experiences, I have perceived that in all cases man must eventually lower, or at least shift, his conceit of attainable felicity; not placing it anywhere in the intellect or the fancy; but in the wife, the heart, the bed, the table, the saddle, the fire-side, the country . . . [p. 349]

The lower intensity of this insight, in comparison with some of the moments of tragic recognition elsewhere in the book, is well signaled by the charming frivolity of the remainder of the paragraph:

> . . . now that I have perceived all this, I am ready to squeeze case eternally. In visions of the night, I saw long rows of angels in paradise, each with his hands in a jar of spermaceti.

Such a comic method of dealing with the issue that determines Ahab's death and Ishmael's survival is far too easy. The affirmation of a heterosexual bourgeois ideal of domestic tranquility, nineteenth-century style, which follows immediately after, can hardly be reconciled with the intensely homosexual imagery of the description of squeezing sperm (pp. 348–349).[26] At the level of plot Ishmael survives by sheer magic: "The unharming sharks, they glided by as if with padlocks on their mouths; the savage sea-hawks sailed with sheathed beaks" ("Epilogue," p. 470). But perhaps a noumenal transaction making his survival possible is hinted at in the words in Melville's handwriting discovered by Charles Olson on a flyleaf of the volume of his set of Shakespeare containing *Lear, Othello,* and *Hamlet.* Immediately after the full version of the blasphemous invocation (*"Ego non baptizo te in nomine Patris et Filii et Spiritus Sancti—sed in nomine Diaboli"*) Melville added: "madness is undefinable—It & right reason extremes of one,—not the (black art) Goetic but Theurgic magic—seeks converse with the Intelligence, Power, the Angel."[27]

This inscription may have been set down before Melville began work on *Moby-Dick* but it could have provided a starting point for an interpretation of madness that would soften the rigid alternatives of craven surrender to conformity or nihilistic self-destruction. From this point of view one might imagine that Ishmael responded to the angelic element, the ideal aspect of Ahab's quest, but withdrew his allegiance when the arrogant excess of it became clear to him. Olson interprets Melville's note as follows:

> I take "it" to refer to the "madness" of the previous sentence. "Right reason," less familiar to the 20th century, meant more to the last, for in the Kant-Coleridge terminology "right reason" described the highest range of the intelligence and stood in contrast to "understanding." [p. 55]

The allusion to Emerson's faculties of Understanding and Reason is open to question. Melville was ordinarily impatient with this kind of technical philosophizing; furthermore, "right reason" is, if not foreign to the Romantic vocabulary, at least rare in it: the Emersonian phrase is usually "the higher Reason" or simply "Reason."[28] Yet I think Olson's interpretation points in the right direction. Although he has to fill in by hypothesis a considerable hiatus in logic, Melville's meaning does indeed come out something like this: the madness of Ahab leads him to extremes of language and apparent paradoxes, such as invoking the Devil in a travesty of the baptismal service. Yet this is perhaps not so depraved as it might seem; it could represent primarily a repudiation of the sterile, routine correctness of established formulas and decorums. Melville's invocation of the Devil might possibly be taken in the sense of Emerson's "Self-Reliance":

> On my saying, "What have I to do with the sacredness of tradition, if I live wholly from within?" my friend suggested,—"But these impulses may be from below, not from above." I replied, "They do not seem to me to be such; but if I am the Devil's child, I will live then from the Devil." No law can be sacred to me but that of my nature. Good and bad are but names very readily transferable to that or this; the only right is what is after my constitution; the only wrong what is against it.[29]

In view of the various explanations and descriptions of Ahab's madness offered by Ishmael, what can be said in summary about it? In the broadest sense it is a *gran rifiuto:* an utterance of NO! in thunder. But what does it repudiate? What is the target of the ferocious hostility that impels Ahab to "the fiery quest"? Superficially, of course, nothing could be more conspicuous than the focusing of Ahab's rage upon the White Whale. But if we then ask what the whale is, what this majestic symbol means, we are back immediately in the middle of the controversies about the book that have been a conspicuous feature of the American scholarly and critical landscape for more than half a century. Brodhead, for example, declares that "Ahab's adversary is in nature, Pierre's in society" (p. 170), and he amplifies that "the question of whether a principle of malicious evil exists in nature is practically embodied in the question of Moby Dick's nature and intentions" (p. 185). On the other hand, Henry A.

Murray declared twenty-five years ago that "Melville's target in *Moby-Dick* was the upper-middle-class culture of his time," or (more fully) "the dominant ideology, that peculiar compound of puritanism and materialism, of rationalism and commercialism, of shallow, blatant optimism and technology, which proved so crushing to creative evolutions in religion, art, and life."[30]

Staats's reading of *Moby-Dick*, combining neo-Marxism and neo-Freudianism, is somewhat similar; it maintains that Ahab's hostility is directed against "culture based on repressive sublimation," the "repression and exploitation of mankind since Adam" by subjecting the race to the reality principle of enforced labor and social conformity.[31] There is strong support for this general view in the "six-inch chapter" devoted to the mysterious figure of Bulkington (Chapter 23, "The Lee Shore"), which has rightly been called a capsule statement of the theme of the novel.[32] It will be recalled that Bulkington serves briefly as steersman of the *Pequod* when the ship sets sail on Christmas night. He has just returned from a four-year voyage, and for reasons that are never stated has immediately shipped again. Yet his course of action is made to seem portentous. Ishmael says that in refusing to stay ashore, but instead setting forth once again to challenge the boundless ocean (and to encounter an unspecified eventual "ocean-perishing" (p. 98)), Bulkington illustrates "that mortally intolerable truth; that all deep, earnest thinking is but the intrepid effort of the soul to keep the open independence of her sea; while the wildest winds of heaven and earth conspire to cast her on the treacherous, slavish shore. . . ." But "in landlessness alone resides the highest truth, shoreless, indefinite as God . . ." (p. 97).

Here again we encounter the same question: Why is the shore "slavish"? I recognize that the Marxists can set forth the oppressive features of capitalism such as reification; that the Freudians point to repression and neurosis or psychosis as the price we have to pay for civilization; and that Herbert Marcuse has brought these diagnoses up to date as of the later 1960s. It may well be that Melville's power of blackness resulted from his ability to recognize aspects of modern society that Marx and Freud and Marcuse would later analyze and make articulate. Indeed, I think it likely that Melville was seeking to express a not dissimilar denunciation of the society he knew in his marvelous fable of a whaling captain's pursuit of a mythical monster of the deep. But whether we take the story at the moment when Ahab climbs upon the mountainous body of Moby Dick to stab him with a six-inch knife, or near the end, when Ahab has succumbed entirely to monomania and accepts his own death and the deaths of his crew as necessary incidents in a consummation endowed with tragic grandeur, Melville seems to me to be making an excessive demand on the reader. We are not given enough fictional substance concerning Ahab's prior experience to account for his state of mind when he first meets the White Whale. Ahab's distinction, his identity—that which makes Ahab, Ahab—is rooted in his madness. But this madness seems to grow more obscure the more Ishmael tries to explain it: it remains to the end what Ahab calls it, a "nameless, inscrutable, unearthly thing," a "cozening, hidden lord and master, and cruel, remorseless emperor" that sends him "against all natural lovings and longings . . . " on a fatal quest (pp. 444–445).

Moby-Dick has a strong appeal for readers affected by the post-Modernist cult of absurdity that has gained momentum since the Second World War. Michael P. Woolf calls attention in a recent article to the large number of contemporary American novels having an insane hero (or anti-hero, since heroes have largely gone out of style).[33] The world, observes Woolf, has seemed to many writers "an organized madness . . . transformed into a social norm," in which "action is devoid of meaning, heroic or otherwise." Since "heroic action is ostensibly inappropriate," the novelist is impelled to create protagonists who "are, in effect, madmen" because they "refuse to accept the limitations imposed" by their environment (pp. 259–261). The critic lists, among others, Ken Kesey's *One Flew Over the Cuckoo's Nest* (1962), Joseph Heller's *Catch 22* (1961), Joan Didion's *Play It as It Lays* (1970), and Saul Bellow's *Mr. Sammler's Planet* (1969), plus a comic statement of the theme, Philip Roth's *Portnoy's Complaint* (1969)—and we might add Thomas Pynchon's *Gravity's Rainbow* (1973). The protagonists of these novels "pursue some projection of an ideal despite the fact that the pursuit of the ideal is doomed to failure" (pp. 261–262), and their action springs from a state of alienation that conventional observers may regard as insanity.

The resemblances are there, but *Moby-Dick* cannot be transplanted to the twentieth century without leaving behind certain important features. One of these is the sheer personal force with which Melville invests Ahab—the scope and grandeur that warrant comparing him with Shakespearian tragic heroes. The liquidation of individuality in contemporary fiction has destroyed even the possibility of creating a character with comparable distinction. And with the disappearance of tragic grandeur in characters has vanished also the "bold and nervous lofty language" (p. 71) that, for better or worse, Melville regarded as the indispensable medium for communicating the insights attained by Ahab and Ishmael and even Pip when they have experiences transcending commonplace reality. What lies behind the pasteboard mask of appearances in *Moby-Dick* may be the colorless all-color of atheism, but this is still only a possibility, and the ambiguity lends a dimension to Melville's fictive world that makes it, although certainly not simpler or safer, undeniably bigger and more complex than the programmatic meaninglessness of the world of the Keseys and Hellers and Pynchons.

A Textbook of the Genteel Tradition:

HENRY WARD BEECHER'S *NORWOOD*

By the outbreak of the Civil War, the kind of highbrow psychological romance that Hawthorne and Melville had brought to a brilliant consummation in the 1850s had been emphatically rejected both by critics and by the public. At the same time, middlebrow domestic fiction with a religious emphasis in the manner of Susan B. Warner's *The Wide, Wide World* (1850), and the "sensation novel" of crime and physical adventure aimed at an even less literate audience, were gaining more and more readers. During the 1860s the field was dominated by weekly story papers serializing this popular fiction, and the closely related series of dime novels published by the firm of Beadle & Adams and its competitors. Such material was ground out according to formulas in an essentially industrial process; it had little bearing on the development of serious literature.

Toward the end of the decade, however, a new generation of significant writers appeared, the generation of William Dean Howells and Mark Twain and Henry James. In order to understand their work we need to understand the cultural situation in which they served their apprenticeships and discovered their characteristic voices and themes, not only the literary conventions that controlled the genre of the novel but also the system of values that was dominant in American society during the post-War decades—the period that has been variously called the Gilded Age or the Brown Decades or the Era of the Robber Barons. The received values of that society, including its esthetic assumptions, are definitively set forth in a remarkable work of fiction entitled *Norwood; or, Village Life in New England*, by Henry Ward Beecher, that was serialized in 1867 in Robert Bonner's New York *Ledger*. This story paper's circulation of three hundred thousand—equivalent to more than a million today—was by far the largest yet attained by an American periodical. Beecher's book was subsequently brought out in London and in New York, and there were further editions in New York in 1880 (with illustrations) and in 1895. A dramatic version was produced in New York and in Brooklyn, and a burlesque in narrative form (*Gnaw-Wood*) went into two editions.[1]

Norwood has a special relevance to American popular culture, not only because of its wide circulation but also because of the unique status of its author. No one before him, and probably no one since, has been the recognized spokesman for so large a segment of the American people. In a brilliant essay in *Trumpets of Jubilee*, Constance M. Rourke says that at the end of the Civil War he was unequaled as "a dominant public figure" and quotes Lincoln to the effect that Beecher was the greatest among his countrymen.[2] Looking back from our own day, the historian William G. McLoughlin asserts that for some three decades Beecher was "the high priest of American religion. His pulpit was the nation's spiritual center—at least for that vast body of solid, middle-class Protestant citizens who were the heart of the nation."[3] His congregation of two thousand in the Plymouth Congregational Church in the fashionable commuting suburb of Brooklyn Heights was the largest in the United States. Crowds waited every Sunday outside the church in hopes of securing a seat or at least standing room for the service.[4] Beecher's sermons were published weekly and had a large circulation of their own. (For example, the Quaker poet Whittier said in 1860 that he read them "regularly and always with deep interest and general approval";[5] and in the late 1860s Miss Olivia Langdon of Elmira, New York, sent copies of them regularly to her fiancé, a young newspaperman named Samuel L. Clemens.[6]) Beyond the pulpit, Beecher was the most popular orator and lecturer in the country. For decades he delivered an average of more than one hundred lectures each year on non-religious topics. In 1870 his salary was $20,000 a year and he earned an additional $15,000 from writing and lecturing. From 1848 to 1870 he was editor of the *Independent*, an influential non-sectarian religious weekly; and for some years during the 1860s he contributed to the *Ledger* a weekly column in which he discussed everything from child rearing to Italian painting. These columns, like his sermons, were collected from time to time for publication in book form. Bonner, the most astute publisher of his day, paid $30,000 for *Norwood*.[7]

Although Beecher was detested in the South both for his theological liberalism and for his highly publicized support of Lincoln and the Republican party, and was denounced for his "naturalism" by Orestes Brownson, the former Transcendentalist who had been converted to the Roman Catholic faith,[8] he had strong appeal as a man and a thinker to much, possibly most of the Protestant majority in the North and West. William Dean Howells, for example, wrote in the *Atlantic* in 1867 from an amusingly patronizing Boston perspective that Beecher represented "the leading thought and speech of the strong, earnest, self-reliant element—not refined to intellectual subtilty or morbid doubt—which is perhaps the most hopeful element in New York, and which is the beginning of a social rather than a religious regeneration. It is American and good; it has sound sense and wholesome impulses."[9]

Whittier kept portraits of Beecher and Marcus Aurelius hanging in his study, and ranked Beecher with Horace Bushnell for his eminence in "Biblical knowledge, ecclesiastical learning, and intellectual power."[10] When a scandal erupted over the charge by Beecher's close friend Theodore Tilton that Beecher had committed adultery with Tilton's wife—a charge which agitated the nation for months, and which many contemporaries (including Mark Twain) believed to be true despite Beecher's technical exoneration—Whittier declared in a letter: "I have loved Beecher so much! I *cannot* believe him guilty as is charged, and yet it looks very dark."[11] Although Beecher suffered a temporary eclipse, his hold on his followers was such that within a few years he had regained his former popularity.[12]

All this, of course, does not mean that Beecher was a novelist. He did not pretend to be one. In the Preface to *Norwood* he said that Bonner's invitation to write a novel for the *Ledger* caught him by surprise: "A very moderate reader, even, of fictions, I had never studied the mystery of their construction. Plot and counterplot, the

due proportion of parts, the whole machinery of a novel, seemed
hopelessly outside of my studies."[13] The emphasis on "construction"
and "machinery," taken as synonyms for "plot and counterplot," is
significant. Following what was evidently the demotic conception of
the genre, Beecher conceives of his book as being composed of two
distinct elements: a plot constituting the novel proper, and a nostal-
gic depiction of village life in New England in sketches not neces-
sarily integral with the plot. The plot conforms to conventions that
were all but universally taken for granted. It consists of a love story (or
rather for good measure, two love stories, one major, one decidedly
minor). The heroes and heroines are young, handsome, virtuous,
accomplished, and of high social standing, but nothing relates them
particularly to their New England habitat (although Beecher does
ascribe stereotyped Southern traits to the secondary hero, a dashing
Virginian). On the other hand, the depiction of village life is accomplished mainly through characters foreshadowing the local-color
fiction of the next decade. These characters belong to a distinctly
lower social stratum although not the very lowest, for Beecher hints
darkly at the existence of a class of "hangers-on—those who are
ignorant and imbecile, and especially those who, for want of moral
health, have sunk, like sediment, to the bottom." This "bottom of
society," he says, "wages clandestine war with the top" (p. 4). But we
see nothing of the menacing lower class in *Norwood*. The only
character who might by any possibility be assigned to it is a gigantic
black, Pete Sawmill, who is often compared with one kind of animal
or another and has an instinctive rapport with all subhuman crea-
tures. He is, however, devoted to the principal hero and heroine, and
saves the hero's life when he is wounded and taken prisoner at the
battle of Gettysburg.

    The distinction between high and low characters was at least as
ancient as Greek tragedy. It is of course conspicuous in Shakespeare,
and had survived through the centuries in European literature. By the
nineteenth century the distinction was on the point of collapsing
under the pressure of social changes, especially in America, but in the
1850s and 1860s it still prevailed in popular fiction as an apparently
unchangeable vestige of the past. Even Howells, who as an editor of
the *Atlantic Monthly* was obliged by the magazine's strong identification with the Republican party to say what was to be said in favor of

Beecher, had to acknowledge that "All the genteel and grammatical people in Norwood are somewhat insipid. . . ."[14] And the reviewer for the *New Englander*, also favorably disposed toward Beecher, thought that the "cultivated people" were "stilted and soaring," "too heavily Beechered." "They are skillfully draped and masked," he noted, "but through all the dominoes one sees the glistening of Mr. Beecher's eyes, and hears the tones of Mr. Beecher's voice."[15] J. R. Dennett, in the rigorously highbrow *Nation*, was more emphatic. The principal characters, he declared, "on whom the book, as a novel, depends for existence, are the average heroes and heroines of the magazine war-tales, doing in all faithfulness the old duty of their kind." As a novel, he concludes, *Norwood* is "badly constructed . . . with a well-worn plot, [and] with a too well-known prig for a hero and an equally well-known bit of perfection for a heroine. . . ."[16]

Beecher does not seem to have been conscious of the emptiness of the fictional conventions he was following. Perhaps as a clergyman he was still unable to take novels seriously. In any case, his imagination was not stimulated by the lay figures he was using as hero and heroine. Rose Wentworth, who is born in the fifth chapter of the book, is in part a literary heritage from the women novelists of the 1850s who had made much of child heroines with a strong religious bent. But Rose differs from her fictional ancestors in being allowed a perfectly harmonious and happy childhood, leading in her maturity to an untroubled religious faith. Beecher evidently intends Rose to exemplify the theory of "Christian nurture" propounded by his friend and occasional adversary Horace Bushnell, of Hartford, who was often linked with him in contemporary discussion as an exponent of liberal Christianity.[17]

Barton Cathcart, the hero of *Norwood*, long believes himself rejected by Rose Wentworth according to the time-honored formula of a misunderstanding caused by the failure of a letter to be delivered. But this is of minor consequence in comparison with his ideological function. Just as Beecher uses Rose to illustrate the doctrine of Christian nurture, he uses Barton to exemplify the democratic conception of the career open to talent—for Barton, son of an ordinary farmer, rises to the rank of general in the Union Army before he reaches thirty. It is even more important that Barton serves as a foil for

Rose in his religious experience; he suffers from a severe case of unbelief that must be cured before he is worthy to seek her hand in marriage. In order to drive home these doctrinal points, Beecher submerges the ritualistic courtship in what J. R. Dennett called a "mass of brief essays on scores of subjects" such as he had long been accustomed to put into his newspaper columns.[18] Some of these are forthright authorial intrusions, but most are attributed to Dr. Wentworth, father of the heroine, as he engages in conversations with two friends—Lawyer Bacon, a rationalist and skeptic, and Parson Buell, the orthodox Calvinist minister. The psychological details of Barton Cathcart's struggle with religious doubt are presented in the form of a journal kept by the young man, as well as through conversations with Dr. Wentworth. The crucial question of how religious faith can be validated is also one of the principal topics that Dr. Wentworth expounds for the benefit of his cronies.

This material led Howells to complain of "the ruthlessness with which the author preaches, both in his own person and in that of his characters, spinning out long monologues and colloquies upon morals, religion, and the whole conduct of life."[19] Since no reviewer seems to have felt real enthusiasm for the principal characters of *Norwood*, we may wonder what was the basis for the book's popularity. One answer almost certainly lies in the fact that a large audience less discriminating in taste than most reviewers was enchanted by Beecher's preaching in print just as the throngs who crowded Plymouth Church were enchanted to hear him perform in the flesh. The reviewer for the distinctly middlebrow *Putnam's Magazine* gives us an idea of how such readers felt. He announced his disagreement with critics who had expressed a desire for more "structure, or character-painting, or society-sketches, or adventure, or description" in the book; he rejoiced that it "utters with a large freedom whatever seemed fit, to communicate the author's thought of what is the essence of characteristic New England life, together with any other thoughts and views of the author." As a result, *Norwood* seemed to him "luminous and living throughout with kindly and noble feeling, and with the contagious cheerfulness of a happy nature; it is a thoroughly healthy and healthful book, which can scarcely be read without imparting some of its own genial warmth."[20] *Harper's Monthly*, also aimed at a middlebrow audience, said that *Norwood* failed as a novel because it had no story, but pronounced the author's "large utterances of truth and duty," his "touches of humor and pathos," and his "keen appreciation of the beautiful in art and nature" to be excellent.[21]

For readers who shared these opinions, Barton Cathcart's struggle with doubt must have seemed exemplary. And they must have welcomed Dr. Wentworth's careful explanation that Barton's case is

> "a clear instance of that doubt which is widely sprung up in the track of physical science. It arises from the introduction of a totally new *method* of investigation. It must be met on its own ground. If the distinguishing doctrines of grace have their types and root in nature, as I believe they have, then evidence from that source will reach the trouble. The alphabetic forms of moral truth found at large in the world will serve to teach one at length how to read those clearer manifestations of the divine nature, and of moral government, which are perfectly disclosed only in the life and teachings of our Savior." [p. 268]

The end of Barton's suffering comes at last when he hears a robin's "peculiar song which indicates the absence of its mate." Such a bird's song suggested to another resident of Long Island, Walt Whitman, the strong and delicious word death, but in Beecher's story the bird's expression "of love *for* love" arouses in the hearer "a sense of God's care" and "a conception of *infinite* love" (pp. 271–272). Both Walt Whitman the boy, as he remembers himself, and Barton Cathcart feel "absorbed and almost identified" with a "Universal Presence." It is, however, significant that whereas in Whitman's poem the Universal Presence is that "savage old mother," the ocean, which is also death, Barton Cathcart says his experience exemplifies Dr. Wentworth's contention that "the Bible interprets nature, and that nature nourishes the truths of the Bible, and that they are parts of one development, and in harmony" (p. 272, 274).

The anxiety to maintain at least formal contact with evangelical tradition is closer to Susan Warner's *The Wide, Wide World* than to Whitman. John Humphreys, the young theological student in the earlier novel, had explained to the heroine Ellen Montgomery that flowers reveal "the glory and loveliness of their Creator," which "is written as plainly to me in their delicate painting and sweet

breath and curious structure, as in the very pages of the Bible.
. . ." But Humphreys had added at once, ". . . though no doubt
without the Bible I could not read the flowers."[22] In Beecher's
formulation, seventeen years later, this qualification is dropped, but
he still implies he is not breaking with theological tradition.
Beecher's method is set forth in the following passage from Dr.
Wentworth's conversation:

> "A flower is beautiful; forests and mountains are noble; clouds meas-
> ure the whole scale, from simple beauty to superlative grandeur. But,
> after all, sunlight, as an object of pleasure, of admiration, and even of
> affection, in the sense in which the term is applied to insentient things,
> is far beyond them all. There are no storms or convulsions in it. Its
> waves fill up the universe, but never rage nor utter sound. . . . Out
> of its stillness come all those energies which awaken life upon the globe.
> . . . It is to me the most impressive feature of the world. It is that
> symbol which most nearly represents the universality of God, the
> energy and fruitfulness of Divine power, and its modesty, as well."
> [pp. 234–235]

In implying that insentient things can be objects of affection only in
a metaphorical sense, Beecher seems to concede that there is not an
actual divine presence within physical nature. To this extent he
shares the assumptions of William Paley's *Natural Theology* (1802).
But Beecher's position differs drastically from the rationalist tradition. His argument has nothing to do basically with logic, but invokes the vocabulary of sensibility and emotion.

When Pastor Buell asks Dr. Wentworth, "Do you think a flower,
in and of itself, has any moral meaning?" the doctor answers with an
equivocation:

> "Do you think that words, in and of themselves, have any signifi-
> cation? Words mean whatever they have the power to make us think of
> when we look on them. Flowers mean what sentiment they have the
> power to produce in us. The image which a flower casts upon a
> sensitive plate is simply its own self-form; but, cast upon a more sensi-
> tive human soul, it leaves there not mere form, but feeling, excite-
> ment, suggestion. God gave it power to do that, or it would not have
> done it." [p. 55]

Orestes Brownson asserted that Beecher was attacking Calvin-
ism in order to propagate "pure naturalism." Of course, Brownson
regarded Emerson's naturalism as self-evident.[23] What he had in
mind in both cases was the maneuver of transferring the basis of
religious faith from a divinely inspired Scripture to intuitions derived from contemplating nature. The charge is true with regard to
Emerson, but Beecher's position differs precisely to the extent that
he was determined to maintain his status as a Christian.[24] One is
tempted to conclude that his followers could tolerate almost any
revision of doctrinal substance so long as he conveyed the impression by his fervent tone that nothing essential was being changed.
Beecher's behavior during the long-drawn-out and highly publicized hearing and court proceedings growing out of the Tilton
scandal shows formidable powers of rationalization.[25]

It was the fusion of evangelical Christianity with Transcenden-
talism that George Santayana called the genteel tradition. From the
perspective of the early twentieth century, he declared:

> . . . the very heart of orthodoxy has melted, has absorbed the most
> alien substances, and is ready to bloom into anything that anybody
> finds attractive. [Transcendentalism] . . . is a method which enables
> a man to renovate all his beliefs, scientific and religious, from the inside,
> giving them a new status and interpretation as phases of his own experi-
> ence or imagination; so that he does not seem to himself to reject any-
> thing, and yet is bound to nothing, except to his creative self.[26]

William G. McLoughlin applies this generalization to Beecher's
specific historical situation:

> In *Norwood* the last vestiges of New England Calvinism disappeared
> finally and forever, and in its place the new theology of liberalism at
> last emerged in terms which everyone could understand. . . . to
> middle-class evangelicals like his [Beecher's] parishioners in Brooklyn,
> it was a marvelous revelation of just what they needed and wanted.

For this reason, McLoughlin calls *Norwood* "in many respects a
textbook" of the genteel tradition.[27]

In fairness to Beecher, I should emphasize that these attitudes
represented powerful imperatives of American culture. Beecher did
not originate them; he simply happened to be supremely capable of

articulating them. The focus of infection was the subtle shift that had taken place in the notion of idealism or, as Beecher's interest in phrenology sometimes led him to call it, "ideality." At the outset of Emerson's career in the 1830s he had proclaimed a subtle idealism in protest against the Lockian empiricism of the Unitarian conservatives who dominated the culture of Harvard and Boston. He had used it to attack what he called the materialism and corpse-cold rationalism of the Unitarian establishment. But thirty years later this idealism had grown crude and dogmatic even on Emerson's own lips. Perhaps the process had been helped along by his gradual loss of memory that set in during the 1860s and became almost complete before his death in 1882; but American culture in general had grown coarse with the unfolding of Manifest Destiny and the demoralizing impact of the Civil War. Charles Eliot Norton noted in his journal in 1873 when he was returning from Europe on the same ship with Emerson:

> Emerson was the greatest talker in the ship's company. He talked with all men, and yet was fresh and zealous for talk at night. His serene sweetness, the pure whiteness of his soul, the reflection of his soul in his face, were never more apparent to me; but never before in intercourse with him had I been so impressed with the limits of his mind. His optimistic philosophy has hardened into a creed, with the usual effects of a creed in closing the avenues of truth. He can accept nothing as fact that tells against his dogma. His optimism becomes a bigotry and, though of a nobler type than the common American conceit of the pre-eminent excellence of American things as they are, has hardly less of the quality of fatalism. To him this is the best of all possible worlds, and the best of all possible times. He refuses to believe in disorder or evil. Order is the absolute law; disorder is but a phenomenon; good is absolute, evil but good in the making.[28]

But this was an Emerson long past his prime. As Stephen E. Whicher demonstrates in *Freedom and Fate*, the authentic Emersonian doctrine of symbolic perception incorporated a dialectic element that could approach a tragic insight.[29] There was an Emerson who declared that "Nature is no sentimentalist,—does not cosset or pamper us," who was aware of "hints of ferocity in the interiors of nature," and who knew that "No picture of life can have any veracity that does not admit the odious facts."[30] Beecher's idealism moves in exactly the opposite direction. He degrades the doctrine of symbolic perception by abandoning its intellectual component.[31] Whereas Emerson's conception of Reason was derived in large part from the Moral Sense of the Scottish common-sense realists, Beecher identifies the power to interpret nature with the heart, that featureless label for the yearnings of popular culture in the middle decades of the century. Dr. Wentworth, in *Norwood*, asserts:

> "I have no theory. I have an irregular and fitful conviction that there are great truths of the affections seeking an inlet upon men, which flow from God, and which reach men, rightly sensitive, through the doings and appearances of what we call Nature." [p. 52]

Even in this brief quotation one can recognize both a general debt to Emerson's doctrine that natural facts symbolize spiritual facts, and the blurring and distortion to which his epistemology has been subjected. In its original form, Emerson's distinction between mundane Understanding and Reason had maintained that this higher faculty, usually identified with intuition or Imagination, could perceive truths inaccessible to empirical observation and analysis. But Beecher's redefinition of the way man arrives at ultimate truth through contemplation of nature leaves no function for the Imagination; in his view, ordinary Understanding needs no help except "the affections" in penetrating the mysteries of the realm of spirit. Thus he transforms the Transcendental doctrine that the universe perceived by the senses is "a metaphor of the human mind," a "language for the beings and changes of the inward creation,"[32] into the solipsistic notion that what we eagerly desire to be true, is true.

Beecher's concentration on the reassuring messages transmitted by nature had implications for esthetic theory that he was eager to develop. He introduced into *Norwood* an artist named Frank Esel (a canting allusion to "easel") for the purpose of expounding the author's views of art. The topic is relevant to the present inquiry because the term "realism" was in the process of being imported from the vocabulary of painting into that of literary criticism.[33] Thus Esel's rather incoherent musings touch on the main issues that would be canvassed in the debate over realism in literature which would break out in the 1870s. He sets down in a letter a satiric thumbnail sketch of a New York artist who has come to Norwood on a summer sketching trip:

> There is one big fellow here whom I found sitting before a most charming view, busily at work painting a board fence, with pig-weed growing by it, and talking about conscience, and painting only "what he sees." He has been working a week, and several knot-holes are yet to be painted in his fence. I looked over his sketches last night. He has one toad, a clump of plantain leaves, a pile of wood, and a heap of stones. I asked him why he selected such subjects. He said "that there could be no true success without humility. An artist must paint what he sees. Nothing in nature is to be despised. He should begin at the bottom and work his way up. It is man's arrogance and egotism that lead him to disdain these lower forms of existence."[34]

Esel's answer to the big fellow from New York evidently has Beecher's approval: "I replied 'Art is not, like science, to investigate and register all natural objects and phenomena. It attempts to work out its end solely by the use of the beautiful, and the artist is to select only such things as are beautiful'" (pp. 193–194).

This immense oversimplification rests on philosophical principles that are the opposite of realistic. At the same time, Beecher's assumption that "natural objects" or "things" are in themselves either beautiful or not beautiful contradicts the Emersonian belief expressed in the reference, in his essay *Nature,* to the use of the microscope: "There is no object so foul that intense light will not make beautiful [*sic*]."[35] It is also instructive to contrast Beecher's doctrine with Whitman's:

> . . . limitless are leaves stiff or drooping in the fields,
> And brown ants in the little wells beneath them,
> And mossy scabs of the worm fence, heap'd stones, elder, mullein
> and poke-weed.

The root of the matter is that despite the Transcendental flourishes in Beecher's remarks about nature, his esthetic doctrine rested solidly on the common-sense epistemology that had been taught in virtually all American colleges for decades. It would persist in the thinking of genteel literary critics as the unexamined premise that fiction ought not to invite the reader to identify himself with characters he

would not care to associate with in real life. The logical extension of
such a doctrine would have limited the novelist's choice of protagonists to the refined and idealized ladies and gentlemen that the
reviewers of *Norwood* declared to be lifeless automatons.

Interestingly enough, Beecher himself recognized that the notion
of ideality, applied to characters, had unfortunate consequences.
The Virginian, Tom Heywood (in this matter expressing the author's own opinion), writes to his brother from the village of
Norwood:

> "I am studying this Yankee people with the utmost zest. Of course,
> many of them are like our own folks. Cultivated people are always
> more or less alike the world over. On that very account one studies
> the middle and lower classes for distinctive characters, as there, if any-
> where, is apt to be found originality and eccentricity." [p. 282]

Heywood's assumption of superiority to the middle and lower classes,
like that of Beecher himself, lies below the threshold of consciousness. The writer was not aware of the dilemma that the attitudes
expressed in Heywood's letter created for the novel as a genre. If
cultivated characters (that is, those who were in contact with the
realm of the ideal) were devoid of interest precisely because of their
approach to perfection of character and manners, then a novelist
trying to hold the attention of his readers would presumably give
prominence to the eccentric and picturesque middle and lower
classes. But literary convention decreed that members of these classes
could not be taken seriously—that is, endowed with freedom; they
must be either comic or pathetic. As long as such assumptions
prevailed, low characters could be introduced into fiction only as
Heywood views them, from the perspective of a higher class to
which writer and reader were also assumed to belong. This is exactly
the formula of the local-color fiction that would fill the magazines
for the next twenty or thirty years. Such writing has a dissociated
sensibility built into it in the form of its implicit class bias. On the
surface it purports to endorse a democratic ideology, but in fact it
expresses an aristocratic one.

This is certainly the case with *Norwood*. Beecher cannot write a
serious work of fiction because the conventions of the novel as they
were delivered to him by a decadent and vulgarized high culture

can not be reconciled with his genuine interest in the common people of his New England village. It is worth noticing, however, that despite his helplessness in actual practice, he sets forth in his Preface the ideology that was already visible on the horizon of the future and would be realized gradually during the remaining decades of the nineteenth century. In explaining how he nerved himself to undertake the writing of a novel even though he had little familiarity with the genre, he says:

> I reflected that any real human experience was intrinsically interesting; that the life of a humble family for a single day, even if not told as skilfully as Wordsworth sung the humble aspects of the natural world, or as minutely faithful as Crabbe depicted English village-life, could hardly fail to win some interest. [p. v]

And further: "The habit of looking upon men as children of God, and heirs of immortality, can hardly fail to clothe the simplest and most common elements of daily life with importance, and even with dignity" (idem). But Beecher can not carry out in practice the doctrine he states at a high level of abstraction. He does not manage to invest the common people in *Norwood* with the intrinsic dignity that Erich Auerbach points to in discussing the story of Peter's denial of Jesus in the gospel of Mark.[36]

Nevertheless, because Beecher is not trying to idealize the uncultivated folk in his story, he does not feel obliged to follow any supposed rules in portraying them. He obviously enjoys this freedom and most reviewers approved of what Howells called his "felicity in expressing the flavor and color of New England life in the talk of . . . such people. . . ."[37] Beecher had an unusually good ear for dialect, and an unusual ability to render the effect of it in print. J. R. Dennett, the critic for the *Nation* who was so severe in his criticism of Beecher's upper-class characters, conceded handsomely that he "stands high among the dozen or so of prose writers who have attempted to delineate the Yankee," and ranked him as a humorist above Thomas Haliburton, creator of Sam Slick.[38] *Harper's Monthly* said the book was "admirable" as a "series of pictures of New England life and character."[39] In Howells's notice in the *Atlantic* he called particular attention to the hostler Hiram Beers, whom he described as "the marked success of the book, . . . exceptionally well-handled. . . . " He quotes several hundred words from the talk of this "pure and simple Yankee" in whom, as he says, "divine grace has compromised with the sinful love of fast horses, and who commonly finds so much to engage him in the teams of the worshippers outside the church on Sundays, that he is apt to be delinquent at the services within."[40]

Technically low characters, usually rural, had attracted the attention of novelists for a long time. As early as 1850, reviewers of *The Wide, Wide World* who found the "moral and religious reflections" tedious had enjoyed the "racy peculiarities" of Aunt Fortune and Mr. Van Brunt.[41] The *Literary World* asserted that Aunt Fortune was "sketched with considerable humor, and several scenes of rude country life are presented in a very agreeable style."[42] Fifteen years later Henry James still remembered the low characters in this book.[43] The subtitle of *Uncle Tom's Cabin* (1852) is "Life Among the Lowly," and although Mrs. Stowe's principal effects were of pathos, she showed remarkable comic gifts as well—for example, in the scene (Chapter 6) in which slaves on the Shelby plantation contrive to delay the slave-trader Haley in his pursuit of Eliza by pretended difficulties in catching up his horse. In 1862, E. P. Whipple had found the hero of Mrs. Stowe's *The Pearl of Orr's Island* poorly drawn but praised the minor characters as good examples of "New-England and human character." He was particularly impressed by one old maid notable for her local "idiom, her crusty manners, and her eccentricities." In fact, the portrayal of "unsophisticated New England life" was authentic enough to reconcile Whipple to the genteel heroine: "This foundation of the story in palpable realities, which every Yankee recognizes as true the moment they are presented to his eye, enables the writer to develop the ideal character of Mara Lincoln, the heroine of the book, without giving any sensible shock to the prosaic mind."[44]

Non-genteel characters, then, not only possessed color and interest, but constituted palpable realities in contrast with the poetic unreality of idealized genteel characters. Nevertheless, low characters could not be allowed to occupy the center of the stage. Howells's reference to Beecher's portrayal of "the quaint ins and outs of Yankee nature" reveals that even he was still not entirely free from conventional class-consciousness, for "quaint," like "picturesque,"

is a patronizing epithet. Auerbach's definition of realism in literature calls for the serious depiction of commonplace characters.[45] But
when Howells began his career as a writer, no American novelist
had managed to treat low characters seriously without idealizing
them (as Cooper had idealized Natty Bumppo, and Mrs. Stowe had
idealized Uncle Tom and Dred).

My argument so far has attempted to illustrate the blighting
effect of the genteel doctrine of ideality on both kinds of characters
allowed by convention in the novel. Idealized straight characters
were deprived of color and spontaneity by the notion that they must
be shown free of blemishes either outer or inner. Low characters
were required to be either comic or pathetic, but in neither case
could they be taken seriously. Cultivated upper-class villains were of
course allowed, but they were obviously unfree, being locked into
their categorical (that is, idealized) wickedness.

Evil not caused by individual wickedness presented another
difficult problem for the novelist. *Norwood* was written only two
years after the end of a war that cost the nation, North and South,
more than a million killed and maimed out of a population of some
thirty-five millions. A writer who took seriously Frank Esel's principle that "the artist is to select only such things as are beautiful"
might well have decided that the War was not an appropriate subject
for representation in fiction. Daniel Aaron's conclusion, after a
thorough study of the depiction of the War in American literature,
that it was essentially "unwritten," suggests that writers generally
found it an awkward topic.[46] A conventional approach could yield
sentimental glorifications of heroic warriors such as the series of
romances by the Southerner John Esten Cooke beginning with
Surry of Eagle's Nest in 1866; or the nascent impulse toward realism
could produce the battle scenes of John W. De Forest's *Miss Ravenel's
Conversion from Secession to Loyalty* (1867), which did not affect
the author's comfortable assumption that secession was the wicked
outcome of a social system corrupted by slavery. Beecher adopts an
entirely different approach. His role as a semi-official adviser to
President Lincoln and American spokesman in Great Britain rendered it all but inevitable that he would make a more frontal attack
on the moral problem of the War. With a magisterial disregard for
plausibility he brings all his genteel lovers to the battlefield of

Gettysburg, kills the elegant and respected Southerner who has chosen the Confederate cause, subjects Barton (now General) Cathcart to grievous wounds, and has him captured and rescued by Pete Sawmill in a thrilling exploit.

Beecher describes the battle in detail on the basis of the dispatches of two celebrated newspaper correspondents (given due credit) and moralizes the spectacle of the corpse-strewn battlefield in his most eloquent pulpit manner. "Could a pitiful God look down through the air on such a scene," asks the narrative voice, "and not fill it with his sympathy . . .?" (p. 493). Without attempting a detailed examination of Beecher's confusing answers to his own question, we may pass quickly to the relentless optimism of his conclusion:

> Death is but the prophet of life. . . . What if twenty thousand wounded men lie groaning here? It is the price of a nation's life! The instruments of their great conflict were carnal, but its fruits spiritual.
> War ploughed the fields of Gettysburg, and planted its furrows with men. But, though the seed was blood, the harvest shall be peace, concord, liberty, and universal intelligence. For every groan here, a hundred elsewhere ceased. For every death now, a thousand lives shall be happier. Individuals suffered; the nation revived!
> Shine on, O Sun! that beholdest evermore the future! Thou wilt not, glorious Eye of Hope,—ever looking at the ends,—be veiled or mourn because the ways are rough through which God sends universal blessings! [pp. 493-494]

The rhetorical device Beecher employs is like a zoom lens, withdrawing writer and reader with magical speed from a close-up of the battlefield to a point of almost cosmic remoteness. To do Beecher justice, we should remember that he has completely disregarded the idealistic canon of beauty by describing the primitive field hospitals in grisly detail. He has even had Rose Wentworth complete an amputation (successfully of course) when the surgeon she is assisting is killed by a stray bullet. And his determination to reach a reassuring conclusion, no matter how flimsy the grounds may be, is not notably different from that of writers for whom we now cherish more respect. In 1864 Emerson had written to Carlyle: "I shall always respect War hereafter. The cost of life, the dreary havoc of comfort & time are overpaid by the Vistas it opens of Eternal Life, Eternal Law, reconstructing & uplifting Society,—[it] breaks up the old horizon, & we see through the rifts a wider."[47] Even the young Henry James was granted an obscurely cheerful revelation concerning the benefits conferred by the War. In 1865 he declared: "Our civil war has taught us, among so many other valuable lessons, the gross natural blindness—that is, we are bound in reason to believe, the clear spiritual insight—of great popular impulses." And he added: ". . . men's natural deserts are frequently at variance with their spiritual needs; and they are allowed to execute the divine plan not only by their own petty practices, but on their own petty theories; not only by obedience but by spontaneity. We are very apt to do small things in God's name, but God does great things in ours."[48] Whitman's exaltation of the War is well known, and even Mark Twain called it "ennobling," asserting that it left the nation "strong, pure, clean, ambitious, impressionable."[49]

In other words, Beecher's crude optimism reflects the mood of public discourse in the North after the War. American culture in the Gilded Age could not accommodate a tragic vision. Ursula Brumm has made this point well:

> The ideal of the American is basically the innocent man, the new Adam of the new world, who lives in the best of all nations, a nation and a society he has shaped according to his own ideals. In this society he devotes himself to peaceful endeavors and the pursuit of happiness, firmly convinced that he is fairly entitled to a happy life.[50]

That is, in assimilating himself to the myth of Adam, the American does not adopt the Calvinist emphasis on our great forefather's guilt, but proclaims himself to be innocent. Yet "The innocent man is at the same time an incomplete man: he must shun experience, for experience brings guilt."

> Tragic consummation . . . demands of its conferee, paradoxically enough, that he be initiated into guilt. . . . He must accept responsibility and punishment as the price for his inner victory in defeat. But for the innocent hero there is no escape. He cannot be reconciled with his situation as a tragedy because he is convinced that he does not deserve his misfortune. The American hero freed of the conviction

of total depravity is unable to deal with the problem of evil in the
world.[51]

Here lay the most seriously paralyzing effect of the genteel cult
of ideality on American fiction. An idealized hero or heroine conceived according to the conventions that had established themselves
in the American novel could figure in melodrama but not in tragedy,
for such a character—a Rose Wentworth or a Barton Cathcart—was
taken to be perfectly innocent. A hundred years later, we can see
clearly that the theory and practice of ideality no longer offered
anything usable to the novelist.

William Dean Howells:
THE THEOLOGY OF REALISM

The great popular success of Beecher's *Norwood* indicates that at the end of the Civil War, the novel as a serious genre was on the verge of bankruptcy. The ablest members of the new generation of writers that emerged in the later 1860s—Mark Twain, Henry James, and William Dean Howells—could discover a viable form in which to express themselves only by breaking more or less completely with the past. In the remainder of this study I shall examine the widely various solutions that the three men found for their common problem.

Howells, although his equipment for writing fiction was by no means negligible, was the least gifted of the three; but his career is the most instructive for the historian because it developed most closely in contact with the institutions that sustained the production and distribution of literature in the United States. His literary apprenticeship was served in unusual circumstances. As a political reporter in Columbus, Ohio, he had written a biography of Abraham Lincoln for the campaign of 1860. The victorious Republican party had rewarded his services by appointing him American consul in Venice, then the principal seaport of the Austro-Hungarian empire. Since trade between Austria and the United States was reduced almost to zero by the Civil War, Howells's position as consul was something like an extended Guggenheim fellowship. He was able to complete his scanty education by residence in one of the great repositories of medieval and Renaissance art, and at the same time to engage in experimentation as a writer without risking his means of livelihood.[1] His published work during these years takes on a special interest because he was subject to so few constraints—in effect, none except the guilt he felt because he was avoiding military service.

The course that he followed was all but inevitable: he exploited his experience as a journalist by writing newspaper letters about Italy. It should be remembered that only a year before Howells went to Venice, Hawthorne had explained in the preface to *The Marble Faun* that he had laid his story in Italy because that country constituted a "poetic or fairy precinct" where "actualities would not be so terribly insisted upon, as they are, and must needs be, in America."[2] That is, Italy was material that had an abundance of ideal or literary interest built in. A simple factual report on Italian life might transcend journalism by virtue of this fact alone. In certain important respects Howells's situation paralleled that of another Western newspaper-man who would soon also be busy making his first book out of travel letters he had sent back from the Old World. Howells's *Venetian Life* appeared in 1866 and his *Italian Journeys* in 1867. *The Innocents Abroad,* by Mark Twain, appeared in 1869. Both men were perceived by their contemporaries as Westerners and as humorists, and we can recognize now that both of them were trying to discover a literary form other than the conventional novel for expressing fresh insights into American experience.

The narrator of *Venetian Life* announces in the first chapter that he will not "fatigue" the reader with any affairs of his own, "except as allusion to them may go to illustrate Life in Venice"; and this promise is kept. The fact that the narrator, like Howells, gets married during his stay in Venice is barely mentioned, and then only to introduce a chapter on housekeeping, devoted principally to a portrait of Giovanna the serving-woman.[3] Nevertheless, the narrator reveals significant assumptions about himself as a representative of the United States. In a masterpiece of inadvertent understatement concerning the Civil War, he says his country is committed to "wholesome struggle in the currents" that represent "the motion of the age," whereas Venice seems to him a "lifeless eddy of the world, remote from incentive and sensation" (p. 38). Indeed, the atmosphere of indolence in Venice affects him as if it were a Tennysonian land of lotos-eaters; as he begins to absorb the spirit of the place he is tempted to doubt whether there is any real reason "for telling the whole hard truth of things . . . " (p. 39). These few phrases are enough to show that Howells has internalized the standard nineteenth-century American ideology: the secular faith in Civilization and Progress. This ideology was so pervasive that Howells was hardly aware of it, but so powerful that it became for him a veritable category of perception. In the year of Gettysburg, he was able to remark casually that he came from "a land where every thing, morally and materially, was in good repair . . . " (p. 37).

It is amusing to realize how closely the attitude of this American pilgrim resembles that of the narrator of *The Innocents Abroad*, despite their contrasting attitudes toward religious paintings by Old Masters.[4] The important technical question is where the American confronted by Europe places himself along the dimension of "high" and "low" in the narrative structure. The narrator of *Venetian Life* is obviously an outsider and a man of considerable leisure (his duties as Consul are barely mentioned). He takes it for granted that he and the reader are cultivated Americans; however sympathetic may be his attitude toward the common people of Venice, he is superior to them in social as well as economic status. He does not develop the theme of philistinism for comic effect, as Mark Twain would do in *The Innocents Abroad*. Yet he admits to a lack of familiarity with medieval and Renaissance art, and he evidently thinks of himself as defying literary convention by dealing with everyday life and the common people rather than with churches and palaces and painting, or with historical associations.

Nevertheless, the matter was not so simple as that. Venice, the object of Ruskin's adoration, was drenched with ideal associations; it was inherently a genteel subject.[5] In a favorable review of *Venetian Life* in the *North American Review*, James Russell Lowell praised Howells's delicacy and refinement and expressed his pleasure that "a natural product" like this young writer from the West (natural in the sense that he was without college education or the cultural opportunities of a Bostonian) should be so "cultivated."[6] What this meant was that the self-educated Westerner exhibited the exact mixture of impatience with the decadent features of dominant cultural values, and basic allegiance to these values, that the Harvard professor and editor of the *North American Review* found in himself. In 1859 Lowell had written to Harriet Beecher Stowe as follows, in a letter praising *The Minister's Wooing:*

> My advice is to follow your own instincts,—to stick to nature, and to avoid what people commonly call the "Ideal"; for that, and beauty, and pathos, and success, all lie in the simply natural. We all preach it, from Wordsworth down, and we all, from Wordsworth down, don't practice it. Don't I feel it every day in this weary editorial mill of mine, that there are ten thousand people who can write "ideal" things for one who can see, and feel, and reproduce nature and character?[7]

Howells was also entangled in this ambiguity; no young writer of that generation could fail to be, unless he could by-pass the role of Man of Letters completely by calling himself, as Mark Twain started out by doing, a humorist. And even Mark Twain was not able to free himself entirely. For example, he showed an almost obsessive concern with the topic of decorum. Much of his early writing was burlesque of various aspects of gentility, yet on the other hand many pages of *The Innocents Abroad* attempt to parade the writer's refinement and cultivation.

Venetian Life, like *The Innocents Abroad*, depends to a considerable degree on the existence of a body of earlier writing about the subject which the reader is now invited to perceive as unauthentic. The treasures of Venetian art had been lavishly described, by Ruskin and many others. There was no other obvious reason for writing a book about a city that had long ago lost its economic and political importance. To some extent Howells assumes the role of an iconoclast in his programmatic focus on the common people rather than on the aristocracy, present or past, and the monuments this aristocracy had erected to itself. But even though he has put aside conventional rhapsodies about painting, sculpture, and architecture, and takes little interest in evoking historical associations, he has still not established his independence of conventional sources of ideal interest. The common people he describes are presented as exotic and picturesque; in place of the splendor of mosaics and murals, he simply gives the reader bits of local color such as the old vendor "burning coffee" in the Piazza di San Marco (p. 35). Furthermore, in order to fill the gap in the fictive structure left by the exclusion of upper-class characters, Howells depends heavily on the narrator and to a lesser degree on the narrator's bride; ideal attitudes and associations are introduced into the narrative through the consciousness of these privileged observers. The conventional class system is modified by defining an American type of gentlemanliness that is not hereditary but the fruit of education and sensibility (p. 383); yet the hierarchic principle remains and is indeed still the major structural principle of the fictive world.

When Howells returned to the United States at the end of the Civil War, he found that his letters to newspapers had already earned a modest reputation for him. In particular, those which had appeared in the Boston *Advertiser* had found approval in the Harvard community. Against all probabilities, after Howells had served for a few months on the staff of the newly established *Nation* in New York, this young Westerner with virtually no formal education was invited to become the principal assistant to James T. Fields on the editorial staff of the *Atlantic Monthly*, of which Professor James R. Lowell had been the first editor only a decade earlier. Howells eagerly accepted the invitation as an opportunity to continue his literary apprenticeship. During the next ten years, in addition to his editorial work (he was made editor-in-chief in 1871), he produced seven novels, a book of poems, and several miscellaneous volumes. *A Modern Instance,* signalizing his arrival at full maturity as a novelist, was his first effort to apply fully in practice the theory of literary realism that he had been developing as an American variant of the European variety for more than a decade in collaboration with his colleagues on the staff of the magazine. This theory had taken shape as an attack on accepted practices in fiction. As early as 1870 Howells had praised Bjørnsterne Bjørnson because his work offered "a blissful sense of escape from the jejune inventions and stock repetitions of what really seems a failing art with us."[8] When Howells later called *A Modern Instance* his "strongest" novel he presumably meant that it was his most successful effort to break with tradition.[9] Unfortunately, however, he was unable to carry through his program of revolt to its logical conclusion. After a brilliant beginning and a well-sustained middle section, his adventure in realism collapsed, and in the last ten chapters or so he reverted to a markedly conventional perspective on his materials that found expression in archaic practices in style and characterization. I shall examine *A Modern Instance* in some detail because the contrast between the later and earlier parts of the book might be said to exhibit in capsule form the past and the future of the novel in this country.

Since there is no reason to assume that everyone is familiar with *A Modern Instance*, let me say briefly that the story opens in the small town of Equity, Maine, in the early 1870s. Marcia Gaylord, only child of the town's leading attorney, is passionately in love with Bartley Hubbard, the handsome and clever young editor of the town's weekly newspaper. Squire Gaylord opposes her marrying Bartley because he correctly believes him to be an unprincipled egoist, but the young people elope to Boston, where Bartley wins rapid although (as it turns out) transitory success by sometimes unscrupulous means in the bustling world of urban journalism. The Hubbards make a few social contacts through Bartley's college classmate Ben Halleck, a quiet young man, son of a pious but wealthy wholesale dealer in leather, who is lame from a childhood accident. But the body of the novel is devoted primarily to the moral decay of Bartley Hubbard. After some three years of married life punctuated by quarrels (many of them caused by Marcia's inordinate jealousy) Bartley flees to the West with a considerable sum of money borrowed from Halleck, abandoning his wife and their two-year-old daughter. At this point virtually a new story begins, having for its subject the chaste but intense love of Ben Halleck for Marcia. He goes to South America for two years in the hope of overcoming his passion but returns unchanged. Through an implausible coincidence Ben learns that Bartley has brought a fraudulent suit for divorce in Indiana, and he and his sister Olive accompany Marcia and her father to the small town where the case is to be tried. They arrive in time to have Bartley's suit thrown out of court. Although Squire Gaylord collapses from a stroke in the courtroom, Marcia's counter-suit is successful. The remainder of the novel focuses on debates between Halleck and an elegant lawyer friend of his named Eustace Atherton about the morality of Ben's desire to declare his love for Marcia and make her his wife. These discussions become especially heated after word arrives of Bartley Hubbard's death in the Far West. Ben finally enters the ministry and becomes pastor of a small backwoods church where the leaven or virus of liberalism has not penetrated. Squire Gaylord is dead, and Marcia, who has apparently never realized Ben's feeling for her, is last seen living in total seclusion with her child in the family house in Equity.

In the first part of the story at least, Howells has apparently managed to shift the locus of the ideal values from the mind of a privileged observer to the minds of characters who are faced with moral choices. The realm of the ideal is now immanent, it is clearly to be found within the commonplace reality of everyday life. Yet it has two aspects whose relation to each other is not precisely defined. In its subjective aspect the ideal consists of moral principles that lead characters to choose good rather than evil courses of action. The theory of realism demanded that the novel be scrupulously honest in depicting these choices, and the depiction was not allowed to be merely neutral (if such a thing is conceivable), for it must include representation of the moral valences of the thoughts and actions of the characters. Howells insisted that both "the highest morality and the highest artistry" in fiction depend on its truth "to the motives, the impulses, the principles that shape the life of actual men and women. . . . " In its negative form his doctrine held that "If a novel flatters the passions, and exalts them above the principles, it is poisonous. . . . "[10]

In the objective realm of action and event, the ideal was to be revealed in the operation of impersonal moral forces analogous to natural laws. These laws were to bring about a constant balancing of accounts having much in common with Emerson's doctrine of compensation: immoral acts were to be punished and moral acts rewarded (whether by material good fortune or by inner happiness, or both, is not fully clear). But this prescription was easily fulfilled because the novelist did not need to concern himself with anything except accuracy—that is, care in observation and honesty in representation: every man's experience reveals the inexorable functioning of such principles. Howells denounced "the whole spawn of so-called unmoral romances, which imagine a world where the sins of sense are unvisited by the penalties following, swift or slow, but inexorably sure, in the real world" (p. 47).

Nothing more clearly marks the difference between Howells's

intellectual universe and ours than this touching confidence in "the essential morality of the universe"—a confidence that Henry F. May identifies as the basis of "American innocence" in the period before the First World War.[11] We are not so likely to remember that this optimistic belief, which Howells shared with most of his country-men, also set his and their world apart from that of their grand-fathers and perhaps even of their fathers. The most venerable Chris-tian tradition had held God's ways to be mysterious, beyond human comprehension; and nature, especially human nature, was held to be tainted with original sin which could be overcome only if the unpredictable and incomprehensible grace of God intervened to effect the redemption of His elect. Partly as the result of Howells's exposure in childhood to his father's Swedenborgianism,[12] but partly also as a result of the general tendency of American culture in the nineteenth century, Howells adhered to a vaguely optimistic faith similar to the liberal Christianity of Henry Ward Beecher. His doctrine of literary realism was the logical application of such a faith to the writing of fiction.[13] But the act of imagining characters and a plot which would exemplify the essential morality of the universe was a much more severe test of this doctrine than Beecher's merely discursive and rhetorical statement of it. Although Howells could show convincingly that the break-up of Marcia's and Bartley's marriage as a consequence of their own flaws of character exhibited the operation of a moral law of retribution, the suffering that resulted was distressing to him because it called in question his assumption that the universe was conducted for the sake of human happiness. The unfolding of Howells's plot brought about a crisis of faith for him as well as a technical impasse in the composition of the novel.

This crisis, however, is not in sight as the novel opens, and Howells can give full scope to his remarkable gift for recording manners and speech. In the first scene he mounts an attack on the high-flown sentiments and corresponding vocabulary that were staple fare in conventional love-stories. Bartley has escorted Marcia to a church sociable, and on their return he has a long conversation with her in the parlor of the Gaylord house that he maneuvers from playfulness to serious tones of flirtation. His approach is a brilliant burlesque of exalted love-making in the novels of the period:

Passage A

. . . he began to speak soberly, in a low voice. He spoke of himself; but in application of a lecture which they had lately heard, so that he seemed to be speaking of the lecture. It was on the formation of character, and he told of the processes by which he had formed his own character. They appeared very wonderful to her, and she marvelled at the ease with which he dismissed the frivolity of his recent mood, and was now all seriousness. When he came to speak of the influence of others upon him, she almost trembled with the intensity of her interest. "But of all the women I have known, Marcia," he said, "I believe you have had the strongest influence upon me. I believe you could make me do anything; but you have always influenced me for good; your influence upon me has been ennobling and elevating."

She wished to refuse his praise; but her heart throbbed for bliss and pride in it; her voice dissolved on her lips. They sat in silence; and he took in his the hand that she let hang over the side of her chair. The lamp began to burn low, and she found words to say, "I had better get another," but she did not move.

"No, don't," he said; "I must be going, too. Look at the wick, there, Marcia; it scarcely reaches the oil. In a little while it will not reach it, and the flame will die out. That is the way the ambition to be good and great will die out of me, when my life no longer draws its inspiration from your influence."[14]

This is skillful and amusing, but a dark intimation of trouble in the future underlies Marcia's naïve vulnerability. The events of the story will reveal in Bartley's blandishments much more than the lovers' oaths of tradition. For one thing, he is not entirely cynical; he is to some extent taken in by his own rhetoric. Or perhaps I should say that at least this possibility was present to Howells's mind at the outset. The next morning Bartley will propose marriage to Marcia without really intending to, mainly because he is suffering from indigestion and the accompanying self-pity, and wants her to pamper him emotionally.

When he leaves Marcia's house, however, after the conversation about the formation of character, he is in complete charge of the situation. He carefully refrains from a declaration of love, but Marcia nevertheless allows him to kiss her good-night (apparently for the first time), and we follow him back to the village hotel where he rooms and boards. There he falls into a reverie that is rendered as follows:

Passage B

There were many things about his relations with Marcia Gaylord which were calculated to give Bartley satisfaction. She was, without question, the prettiest girl in the place, and she had more style than any other girl began to have. He liked to go into a room with Marcia Gaylord; it was some pleasure. Marcia was a lady; she had a good education; she had been away two years at school; and, when she came back at the end of the second winter, he knew that she had fallen in love with him at sight. He believed that he could time it to a second. He remembered how he had looked up at her as he passed, and she had reddened, and tried to turn away from the window as if she had not seen him. Bartley was still free as air; but if he could once make up his mind to settle down in a hole like Equity, he could have her by turning his hand. Of course she had her drawbacks, like everybody. She was proud, and she would be jealous; but, with all her pride and her distance, she had let him see that she liked him; and with not a word on his part that any one could hold him to. [pp. 14-15]

"She had more style than any other girl began to have" is to be understood as the report of a remark Bartley makes to himself which originally had the form, "She's got more style than any other girl begins to have." Although the report uses the past tense, it implies a present tense. Later critics would call this kind of writing "represented discourse"—a third-person paraphrase of a putative first-person interior monologue representing a dramatized character's attitudes, and often also the character's own recognizable choice of idiom.[15] The device is at least as old as Chaucer (the passage about the Monk in the Prologue to the *Canterbury Tales*—

> What sholde he studie, and make him-selven wood,
> Upon a book in cloistre alwey to poure?

is a good example of it), but it had been first used systematically in the nineteenth century by French writers, especially Flaubert. Represented discourse makes the advantages of the dramatic soliloquy available in the novel. It seems to place the reader inside a character's mind without the intervention of the author, revealing details of mental processes not accessible to any other character. At the same time, the device allows the writer (as Howells does here) to communicate very forcefully his own evaluation of the character's thoughts. The effect of this passage is to impose on the reader a strongly negative view of Bartley Hubbard, who is made to seem definitely low and vulgar. His feeling about Marcia is at the opposite extreme from what was expected of the protagonist in a love-story. Nothing could have signaled more clearly to the reader Howells's intention of abandoning the conventions of the standardized romance. When Bartley observes that "Marcia was a lady," he is using the term as he might have praised the gaits of a horse; it is unmistakably a coarse remark.

Bartley's anomalous social status is one of the most distinctively modern (and American) features of Howells's story. Whereas Marcia's high position in the society of Equity is determined beyond question by the fact that she is Squire Gaylord's daughter,[16] Bartley is an orphan; we know nothing else about his past except that he has been cared for by relatives who sent him to college. A college degree was much more of a distinction in New England in the 1870s than it is a hundred years later; it is no doubt in large part Bartley's education that gives him his sense of superiority to "a hole like Equity." But Howells insists that college has not made Bartley a truly cultivated man. The text of the serial version of *A Modern Instance* contained the following observation (omitted from the version published in book form): "His college training had been purely intellectual; it left his manners and morals untouched, and it seemed not to concern itself with his diction or accent; so far as his thought took shape in words, he thought slangily; and he now reflected that no girl had ever 'gone for him' so before. . . . "[17] Bartley's meditation about Marcia, then, has the effect of casting doubt on his social status and making him seem an unsympathetic character because of his mercenary attitude toward a girl whom the reader is expected to find attractive for her very innocence and defenselessness.

In the sentimental popular fiction that had dominated commercial publishing for twenty years or more when Howells began his career as a novelist, the categories of high and low characters were rigidly defined. It will be recalled that in reviewing Beecher's novel *Norwood* in 1868, Howells had referred casually to the "genteel and grammatical" characters—all of whom he found "somewhat insipid"—and those of lower status, who represented the "quaint ins and outs of Yankee nature."[18] Now, a decade later, Howells is undertaking to break down the rigid distinction between high and low, grammatical and ungrammatical, by presenting in Bartley Hubbard a protagonist neither clearly high nor clearly low. To blur the conventional categories in this fashion was to move toward realism in Erich Auerbach's sense—"The serious treatment of everyday reality, the rise of more extensive and socially inferior human groups to the position of subject-matter for problematic-existential representation. . . . "[19] Howells's theory of realism was much more moralistic and ideological; he insisted, then and later, that literary realism must exhibit a spirit of "equality running through motive, passion, principle, incident, character, and commanding with the same force [the writer's] interest in the meanest and the noblest, through the mere virtue of their humanity."[20] But his commitment was theoretical rather than a realized fact of the imagination. He had not succeeded in eliminating all traces of class-consciousness from his fiction. Bartley's ambiguous social status influences the attitude of the narrative voice toward him: it seems to be connected with his moral shortcomings.

Bartley's lack of moral principle is illustrated by an incident that forces him to leave Equity. He has been carrying on a casual flirtation with one Hannah Morrison, a girl employed in the printing room of the newspaper he edits, and when a young man who also works on the paper denounces him for misleading the girl, Bartley knocks him down, injuring him seriously although not permanently. Marcia, in the first of a series of jealous rages that will come over her during the course of the novel, breaks her engagement with Bartley, and Squire Gaylord as spokesman for the committee that controls the paper announces it will be discontinued. In this crisis, the reader is placed again inside Bartley's mind. But now the depiction of his mental processes makes almost no use of the method of represented discourse that dominates Passage B:

Passage C

He could rebel against the severity of the condemnation he had fallen under in the eyes of Marcia and her father; he could, in the light of example and usage, laugh at the notion of harm in his behavior to Hannah Morrison; yet he found himself looking at it as a treachery to Marcia. Certainly, she had no right to question his conduct before his engagement. Yet, if he knew that Marcia loved him, and was waiting with life-and-death anxiety for some word of love from him, it was cruelly false to play with another at the passion which was such a tragedy to her. This was the point that, put aside however often, still presented itself, and its recurrence, if he could have known it, was mercy and reprieve from the only source out of which these could come. [p. 67]

The narrative point of view here is not easy to determine. Some kind of debate is going on in Bartley's mind, but it is being reported from a greater distance than is the case in Passage B. "He found himself looking at it as a treachery to Marcia" contains no hint of Bartley's own language; yet ". . . she had no right to question his conduct before his engagement," omitting some such marker as "he thought," verges on represented discourse. In the next sentence, "If Marcia loved him" would be closer to represented discourse than "If he knew that Marcia loved him." Such language as "life-and-death anxiety," "cruelly false," and especially "to play with another at the passion which was such a tragedy to her" is not only implausible as Bartley's phrasing, it represents an entirely different vocabulary from that used by the narrative voice in Passage B or even in the opening sentences of the present Passage C. This is the language of the melodramatic stage. It carries a charge of crude emotion that obliterates nuances and rushes toward hyperbole. As a consequence, whereas in Passage B the slang lends an air of authenticity to the rendering of Bartley's mental process, here the overstrained diction loses its power to convince the reader. Howells is pressing too hard.

The difference between the two passages seems to lie in the writer's attitudes toward his material. In Passage B he is very much in control; indeed, he seems to enjoy Bartley's slang. But here he has lost his poise; his moral indignation takes over, and he all but steps out on the stage in order to speak directly to the reader over the head of his character. This procedure violates a principle that Howells later stated explicitly when, endorsing Turgenev's "dramatic" method of narration, he denounced "the deliberate and impertinent moralizing of Thackeray, the clumsy exegesis of George Eliot, the knowing nods and winks of Charles Reade, the stage-carpentering and lime-lighting of Dickens, even the fine and important analysis of Hawthorne."[21]

But more is involved here than mere consistency in narrative viewpoint. What may seem a finical matter of technique bears directly on the basic principle of Howells's realism. He maintained that the accurate portrayal of ordinary men and women going about their commonplace affairs would convey moral truth to the reader because moral principles are immanent in all experience, are indeed structural features of the actual world. Such principles do not need to be expounded by the novelist; if the material is truthfully depicted they will be apparent to every eye. Passage C, however, presents a rather complicated case because Howells does not clearly and firmly indicate the status of the "point" that Bartley's behavior is "cruelly false." Bartley is being judged according to supposedly self-evident, absolute principles that eliminate moral interest from the situation because they are so completely unable to accommodate subtleties of characterization or evaluation.

The hint of melodrama is made stronger in the gratuitous comment of the narrative voice that certain recurrent events in Bartley's mind represent "mercy and reprieve" for him. Again, the diction surrenders to melodrama by invoking a primitive set of theological assumptions. God, acting through Bartley's rudimentary conscience or moral sense, gives him a chance to repent. But Bartley ignores these merciful promptings and is therefore lost, that is, literally damned. When the action of the story is made to depend on conflicts between such abstractions as these, the fictive world loses its imaginative solidity and becomes merely an allegory of the eternal struggle between right and wrong. There can be no intermediate categories, no fine distinctions. Let me emphasize the difference between this fictive world and that created in Passage B. There, the indirect ironic judgment passed on Bartley gives at least the illusion of subtlety and even flexibility, a recognition that human beings and their motives are complex affairs, for the understanding of which the two categories of right and wrong are not adequate equipment. The minute, almost microscopic depiction of Bartley's feelings in the represented discourse leaves with the reader the impression that he has been observing the operation of a believable human mind immersed in an actual world. There is no way to accommodate divine intervention in such a world without having it recognized as miraculous, and this was no longer acceptable in fiction purporting to represent everyday life. Yet the mercy and reprieve offered in Passage C are covertly supernatural.

Howells returns again and again to Bartley's habit of rationalizing his failures and frustrations by blaming them on someone else. As Bartley drives fifteen miles cross-country to the railway station after leaving Equity for Boston, he reflects upon "the disaster into which his life had fallen."

Passage D
He owned even then that he had committed some follies; but in his sense of Marcia's all-giving love he had risen for once in his life to a conception of self-devotion, and in taking herself from him as she did, she had taken from him the highest incentive he had ever known, and had checked him in his first feeble impulse to do and be all in all for another. It was she who had ruined him. [pp. 102–3]

Bartley can hardly be imagined as thinking that he had "risen for once in his life to a conception of self-devotion,"[22] but the notion that "she had taken from him the highest incentive he had ever known" sounds somewhat more like his way of viewing the situation. The language echoes his self-dramatizing remark in Passage A about her stimulating his ambition to be good and great. And the notion that "It was she who had ruined him" is so outrageous that both the sentiment and the phrasing must be Bartley's.

Howells seizes further opportunities to throw Bartley's depravity into relief by portraying what for a different man might be opportunities for redemption. Let me quote another passage. Defying all decorums, as well as her father's warnings about Bartley, Marcia intercepts him at the railway station and they are married before leaving together for the city. Bartley is deeply touched by Marcia's unexpected appearance. But Howells's description of his state of mind is again highly ambiguous:

Passage E
All that wicked hardness was breaking up within him; he felt it melting drop by drop in his heart. This poor love-tossed soul, this frantic, unguided, reckless girl, was an angel of mercy to him, and in her folly and error a messenger of heavenly peace and hope. [p. 105]

The question of whether the phrase "wicked hardness" is to be ascribed to Bartley or to the narrative voice bears on the degree of self-knowledge and moral insight properly attributable to Bartley. Does he recognize that he has been in a mood of hardness, and that this hardness has been wicked? Is it Bartley, again, or the narrative voice that conceives of Marcia as a "love-tossed soul," a "frantic, unguided, reckless girl"? Even if these questions could be answered with confidence, there would remain the problem of how to interpret "angel of mercy" and "messenger of heavenly peace and hope." How literally should "angel" and "heavenly" be taken? My own opinion is that these phrases describe Marcia not from Bartley's perspective but from that of the narrative voice. Bartley's performance in the opening scene demonstrates his capacity for dramatizing himself, but it is inconceivable that he should talk to himself in the vocabulary of literary neoplatonism. The statement of the narrative voice that Marcia was an angel and messenger to him is to be interpreted theologically; Bartley is receiving again some prompting of divine grace.[23] He is moved to the point of declaring, a moment later, "I am a bad fellow, Marcia," but the impulse ends there; he is not capable of a lasting response to such opportunities.

The interplay of impulses and motives in Bartley's mind is centrally important, of course, in the story of his moral disintegration. It is true that Howells's working title for the book was *The New Medea*, which implies that Marcia is the dominant character.[24] Yet Henry James called the novel a "Yankee *Romola*," implying that its most important character is Bartley, as a counterpart to Tito Melema in George Eliot's novel.[25] We are in fact told remarkably little about what is going on in Marcia's mind. Perhaps the nimbus that had come to surround the heroine in sentimental fiction restrained Howells from invading her mind with realistic analysis. And no doubt Howells had some notion all along of the part she would eventually be called on to play as the object of Ben Halleck's adoration; she could not be made to seem grossly unworthy of it. He keeps her faults—her emotional fixation on her father at the expense of her husband, and the fierce jealousy that seems to be caused by repressed guilt over this emotion—from seeming sordid, as Bartley's selfishness and lack of professional scruple emphatically are. For whatever reason, she is left rather vague in comparison with Bartley and Ben Halleck.

It is Bartley Hubbard's moral disintegration that offers the most promising scope for the application of Howells's theory of realism. As a newspaperman Bartley is a kind of writer; his temptations and vices are of a sort familiar to Howells—and incidentally also to Howells's friend Samuel L. Clemens, who wrote to the author in extravagant admiration of the portrayal of this protagonist. Bartley's behavior raises problems that strain the rigid black-and-white moral categories enshrined in popular culture. He is obviously a scoundrel, yet both Clemens and Howells recognized much of themselves in him. Clemens acknowledged the resemblance in a characteristic letter. "You didn't intend Bartley for me, but he *is* me, just the same & I enjoy him to the utmost uttermost, & without a pang," he declared. "Mrs. Clemens indignantly says he doesn't resemble me— which is all she knows about it."[26] Many years later Howells declared he had drawn "the false scoundrel" from himself.[27] Perhaps this mixture of identification with Bartley and harsh judgment of him accounts for confusion in characterization. Howells prepares an explanation of his behavior in the general circumstances of his life: his being an orphan, for example, and thus the object of a pity that turned readily into over-indulgence. Also, he was apparently given no systematic instruction in religion or morals. But as we have seen (pp. 88–89 above), in portraying Bartley's mental processes in actual crises, Howells sometimes abandons the effort to specify empirical causes, resorting instead to more or less explicitly supernatural forces.

As Bartley approaches the decision to abandon Marcia, the narrative voice asserts that her unjust accusations and irrational jealousy "had hardened his heart . . . past all entreaty." "He was going to have this thing over once for all; he would have no mercy upon himself or upon her; the Devil was in him, and uppermost in him, and the Devil is fierce and proud, and knows how to make many base emotions feel like a just self-respect" (p. 275). The language seems literal. And it certainly does not provide the chain of cause and effect that in moments of theoretical rigor Howells demanded of the practitioner of literary realism.[28] In fact, this doctrine, taken as identical with the requirement of "honesty" on the part of the novelist, proves upon close examination to be simply a positivistic fantasy. Howells had been misled by the optimistic jargon of popular science, in a period when enlightened laymen were confident that the psychologists would soon be able to write the equations for the chemical reactions in the cells of the cerebral cortex that were the demonstrable causes of any given mental event. Yet Howells was also suspicious of elaborate metaphors and symbols in fiction. It is not surprising that in situations of conceptual strain he fell back almost inadvertently upon a buried reservoir of notions belonging to a primitive stage of Christian theology, when it was still impregnated with ideas and images derived from magic and folklore.

Daniel Defoe had brought to the novel in its embryonic phase a rich background of beliefs, Catholic and Protestant, about ministering angels as well as the unceasing activities of the Devil and his legions.[29] But angels began to disappear from serious fiction early in the nineteenth century, and by the 1850s they are glimpsed but rarely even in intensely pious best-sellers such as *The Lamplighter*. The Devil showed greater powers of survival. The common-law form of indictment for crime in England included until quite recently the formula, "being prompted and instigated by the Devil."[30] Although Fenimore Cooper is little interested in gothic flourishes, Natty Bumppo takes this orthodox view for granted. In *The Pioneers* he tells Hiram Doolittle: "You may be formed in the image of the Maker, but Satan dwells in your heart."[31] The "tempter of souls" figures prominently in *The Scarlet Letter*, although this may of course be an aspect of Hawthorne's creation of an atmosphere of historical authenticity.[32] In Rebecca Harding Davis's *Dallas Galbraith* (1868) the "demons" of anger and desire for revenge tempt the protagonist to kill an enemy, but he manages to restrain himself.[33]

Some of Howells's references to the Devil can be given a metaphorical rather than a literal interpretation. Asked point-blank whether he believed in the existence of a personal Devil, he would probably have replied that he found such a notion a useful figure of speech; and this would have been accurate with regard to his conscious thought. He wrote in 1874, for example, in a review of Christopher P. Cranch's *Satan: A Libretto*:

> The tendency of modern liberalism to ignore the chief of the fallen angels has been one of the most painful spectacles which conservative theologians have had to contemplate; . . . whether we call him Devil or call him Disorder, we still have the same old serpent among us for all practical purposes.[34]

In the same vein, many years later Howells asserted that mankind would one day "hold selfish power in politics, in art, in religion, for the devil that it is. . . ."[35] Nevertheless, the references to demons, angels, and vague supernatural agencies as forces motivating Bartley Hubbard in moments of moral crisis are, cumulatively, too substantial to be viewed solely as metaphors. On at least some of these occasions Howells, baffled by his inability to present a convincing account of Bartley's mental processes in the vocabulary of empirical cause and effect, is falling back on beliefs, widely shared by an older generation, which merged with religious doctrines he had been taught in childhood. Perhaps neither he nor his readers could be sure how literally such allusions were to be interpreted. But it is hard to believe that he does not shift occasionally to notions he had consciously abandoned.

In the passages I have quoted from *A Modern Instance*, the two alternative sets of assumptions about the fictive world are associated with two contrasting rhetorical modes. When Howells is writing as an experimental realist within a frame of assumed moral complexity, his diction and syntax tend toward dryness and precision. But as he approaches a crisis, when a character (usually but not always Bartley Hubbard) is forced to make an important moral decision or is to be punished by the operation of supposedly immanent moral laws, a set of polarized, rigid moral categories is brought into play that finds expression in an elevated and turgid diction and an oratorical syntax, the over-all stylistic effect being decidedly archaic. As we have seen, there are some occurrences of this regressive bombastic style in the earlier chapters of the book, but they grow more frequent later, and after Bartley disappears into the Far West, this style dominates the story. I shall illustrate these observations presently by examining a few more passages from the book, but before doing so I should like to cast at least a glance at the problem of causation. What was the reason for the change?

The answer may be hidden somewhere in the immense Howells archive in Houghton Library; but, so far as I know, no one has made a systematic effort to find it there. Kenneth S. Lynn, who has worked in the Howells Papers, says that the latter part of the novel shows "how full of guilt Howells had been—and in some less overt sense still was."[36] This is undoubtedly true, but one would like to have more details. My conjecture runs somewhat as follows: following the determinism latent in his insistence upon "causes and effects," Howells set out to show how the breakdown of a marriage results from "the undisciplined character" of both husband and wife.[37] The undisciplined characters of Marcia and Bartley Hubbard are in turn to be accounted for by specified forces of heredity (in Marcia's case; Bartley's parents are never mentioned) and environment (Squire Gaylord's indulgence of Marcia; the lack of religious training in childhood for either Marcia or Bartley). The novel develops through three-fourths of its course as an unfolding of the consequences that follow from these premises. As Bartley Hubbard's flight into the anonymous West draws near, the fictive universe comes to present more and more clearly the aspect of an unbreakable sequence of causes and effects. Bartley and Marcia seem to have been hopelessly (and therefore, it might seem, blamelessly) doomed from the start— as if by a Calvinistic inevitability without the alleviating force of divine grace.

This prospect imposed severe conceptual strain on Howells. He could not bring himself to accept so bleak a philosophy. I believe the strain was so great that in conjunction with the baffling illness of his daughter Winifred it caused the "breakdown" (a serious collapse lasting several weeks) suffered by the writer after he had composed 1,466 manuscript pages—approximately the first thirty-one chapters of the novel. He apparently forced himself to resume work on the book before he had fully recovered in order to meet publishing deadlines, for the story had already begun to appear in installments. As a result, in the last nine or ten chapters the anxieties that are revealed only occasionally in the earlier part of the novel take control. Howells himself admitted that "the result seems to lack texture."[38] The style reveals that he had undergone a temporary regression according to the pattern described by Erik H. Erikson:

The child internalizes into the super-ego most of all the prohibitions emanating from the social structure—prohibitions furthermore, which are perceived and accepted with the limited cognitive means of early childhood and are preserved throughout life with a primitive sado-masochism inherent in man's inborn moralistic proclivities. (These are aggravated, of course, in cultures counting heavily on guilt as an inner governor.) This internalized infantile moralism becomes isolated from further experience, wherefore man is always ready to regress to and to fall back on a punitive attitude which not only helps him to re-repress his own drives but which also encourages him to treat others with a righteous and often ferocious contempt, quite out of tune with his more advanced insights.[39]

In *A Modern Instance,* these attitudes are expressed in a drastic structural maneuver. When Bartley Hubbard flees westward, Howells moves Ben Halleck into the vacated role of protagonist and the concluding section becomes in effect a different story. Whereas down to the end of Chapter 31 the novel deals with the disintegration of Bartley and the break-up of his and Marcia's marriage, the final section focuses attention on the relationship between Marcia and Ben Halleck. Critics have noticed a curious similarity-in-difference between the two protagonists, as if Howells saw in them a polar contrast. They have the same initials; they are in love, each in his own fashion, with the same woman. Marcia herself makes several remarks indicating that the two men are in some way linked in her mind. Yet whereas Bartley is handsome and well formed, Ben has "a plain, common face" (p. 240) and walks with a limp.

And they are even more different in character than in physique. Whereas Bartley, as Halleck exclaims at one point, "has no more moral nature than a baseball" (p. 170), Ben has an excruciatingly acute moral sense. In order to develop the contrast Howells must subject Halleck to a strong temptation that is not in itself degrading. He does so by causing Halleck to feel a "passion" for Marcia (p. 319) that is evil only because she is married to another man. But since she is married, Howells considers it to be little short of monstrous. The lawyer Atherton, speaking (in my opinion) for the author, maintains first that it is wicked for Halleck to reveal his love to Marcia after her husband has deserted her but while he is not known to be

dead. Then, when news has come of Bartley's death, Atherton con-
tends that Halleck still must not speak because the relation will
always be tainted by the fact that he loved Marcia while her husband
was alive. The lawyer's moralizing is the more egregious in view of
the fact that, as he himself points out, Marcia's attention is fixed on
Bartley and she shows no signs of interest in Halleck as a lover.
Kenneth Lynn calls Atherton's position "hysterically pompous." It
was undoubtedly extreme, even in 1882.[40]

Nevertheless, contemporary critics approved of the "touches of
deeper passion" in the book, especially "the devotion of Ben Halleck
to Marcia"; and the final section was praised as the revelation of a
"higher moral purpose" than had been evident in Howells's earlier
novels.[41] An anonymous reviewer in the *Nation*, while objecting to
the "disagreeable flavor" of the book (probably an allusion to the
fact that it depicts the break-up of a marriage), declared that "as a
work of moral fiction 'A Modern Instance' is unequalled."[42] This
praise indicates once again the malign influence of outworn literary
conventions; for Halleck's "passion" is an unmistakable residue of
exactly the kind of fiction that Howells attacked in his critical
writing. The vocabulary that Atherton and Halleck use is so generalized and at the same time so highly charged with emotion that it
obliterates all fine distinction in the analysis of psychological states.
For example, when Halleck gives his first full account of his suffering to Atherton, he speaks of his "trouble" as "ghastly"; merely
the knowledge of it would "defile" another soul, and it would be a
"profanation," indeed a "sacrilege," for Ben to reveal his inward
turmoil to anyone else. "Pity, rectitude, moral indignation, a blameless life" are of no avail; these virtues can easily become "instru-
ments" for the Devil (p. 291). The issue comes up again in a later
conversation. "Don't you suppose I know how hideous this thing
is?" cries Halleck:

Passage F

". . . when you said that love without marriage was a worse hell
than any marriage without love, you left me without refuge: I had
been trying not to face the truth, but I had to face it then. I came away
in hell, and I have lived in hell ever since. I had tried to think it was
a crazy fancy, and put it on my failing health; I used to make believe

that some morning I should wake and find the illusion gone. I abhorred it from the beginning as I do now; it has been torment to me; and yet somewhere in my lost soul—the blackest depth, I dare say!—this shame has been so sweet,—it is so sweet,—the one sweetness of life—Ah!" [p. 293]

Some of the phrases in this passage create a momentary impression of detailed description or even analysis, but on closer examination they dissolve into the same featureless abstractions that characterize all references to the emotion that overwhelms Ben. "Hell" is not a determinate place in a cosmography like Dante's, nor can "the blackest depth" of Ben's "lost soul" be located on an imaginable map of the psyche. The "shame" that is superlatively "sweet," indeed is "the one sweetness of life," is unquestionably sexual in origin, but this is as close as we ever get to that important fact.[43]

Atherton's language and syntax are indistinguishable from Ben's. An example from a later conversation will serve:

Passage G

". . . how will you ask her, if she's still a wife, to get a divorce and then marry you? How will you suggest that to a woman whose constancy to her mistake has made her sacred to you?" Halleck seemed about to answer; but he only panted, dry-lipped and open-mouthed, and Atherton continued: "You would have to corrupt her soul first. . . . I don't know what change you've made in yourself during these two years; you look like a desperate and defeated man, but you don't look like *that*. You don't *look* like one of those scoundrels who lure women from their duty, ruin homes, and destroy society, not in the old libertine fashion in which the seducer had at least the grace to risk his life, but safely, smoothly, under the shelter of our infamous laws. Have you really come back here to give your father's honest name, and the example of a man of your blameless life, in support of conditions that tempt people to marry with a mental reservation, and that weaken every marriage bond with the guilty hope of escape whenever a fickle mind, or secret lust, or wicked will may dictate? Have you come to join yourself to those miserable specters who go shrinking through the world, afraid of their own past, and anxious to hide it from those they hold dear; or do you propose to defy the world, to help form within it the community of outcasts with whom shame is not shame,

nor dishonor, dishonor? How will you like the society of those un-
certain men, those certain women?" [pp. 317–18]

Scoundrels luring women from their duty and ruining homes, a
seducer risking his life (as Lovelace does in *Clarissa?*), secret lusts,
and guilty specters shrinking through the world are not images one
would expect to find in Howells's fictive universe. They belong to
the detritus of sentimental popular fiction. But I believe these phrases
are set down here without irony; they register the writer's retreat
from the hard-won sophistication of his original plan.

Later, when Ben receives by accident a newspaper containing
the notice of Bartley Hubbard's suit for divorce out in Indiana, his
struggle against the temptation to conceal this information from
Marcia is described in less highly colored language but with a
similar effect of abstraction and oratorical elevation:

Passage H

Why should he break the peace she had found, and destroy her last
sad illusion? Why should he not spare her the knowledge of this final
wrong, and let the merciful injustice accomplish itself? The questions
seemed scarcely to have any personal concern for Halleck; his temptation wore a heavenly aspect. It softly pleaded with him to forbear,
like something outside of himself. It was when he began to resist it
that he found it the breath in his nostrils, the blood in his veins. Then
the mask dropped, and the enemy of souls put forth his power against
this weak spirit, enfeebled by long strife and defeat already acknowledged.

At the end Halleck opened his door, and called "Olive, Olive!" in
a voice that thrilled the girl with strange alarm where she sat in her
own room. She came running, and found him clinging to his doorpost,
pale and tremulous. "I want you—want you to help me," he gasped.
"I want to show you something—Look here!" [pp. 323–24]

The temptation which wears a heavenly aspect and then drops
its mask to reveal itself as "the enemy of souls" must be Ben's
passion for Marcia. Howells stays so far away from psychological
fact that only on reflection does the reader become aware that Ben's
temptation or passion is a kind of blur in which sexual desire
merges with every kind of instinctual or unconscious impulse, and
these in turn are equated with hell and with the Devil. This is a

conventional early nineteenth-century linkage; indeed, the notion of the ideal in that period is best understood as referring to whatever parts of the psyche are free of contamination from the unconscious. Such a two-valued system of classification accounts for the emphatic terminology of melodrama—the numerous pairs of antonyms about which the rhetoric is organized, such as good/bad, spirit/flesh, soul/body, light/darkness, and so on. We leave the realm of empirical events altogether and enter a manichean universe in which vast cosmic forces war with each other.

If Ben's assailant is the Devil, then of course no merely human power can withstand the attack. But these are not matters that can be represented within a supposedly realistic novel. The climax of the struggle is simply a gap in the narrative at the end of which Ben emerges "pale and tremulous" but victorious. How has this victory been achieved? One can hardly avoid the conjecture that for Howells's imagination, Ben was rescued by divine grace, by a miracle. Moral crises are resolved in this fashion repeatedly in popular novels of the 1850s and 1860s, but Howells evidently felt an inhibition against making such supernatural forces explicit; they were not an acceptable part of the fictive universe for the segment of the reading public represented by the subscription list of *Century* magazine in 1882.

I suggest that in the absence of an explicit reference to divine grace, Howells felt obliged to provide a non-supernatural rationale for the redemption of Ben Halleck just as he had provided a non-supernatural rationale for the damnation of Bartley Hubbard. Again he resorts to Atherton as his spokesman. It is true that at the very end of the novel Howells casts some doubt on Atherton's moral authority, but the lawyer strikes me as speaking most of the time with the author's full support.[44] In the last scene of the book, Atherton explains to his wife why Halleck was able to overcome temptation whereas Bartley was destroyed by it. Atherton's wife remarks that Ben was always "naturally" a "good soul," but Atherton is emphatic in correcting her:

Passage I
"The natural goodness doesn't count. The natural man is a wild beast, and his natural goodness is the amiability of a beast basking in the sun when his stomach is full. The Hubbards were full of natural

goodness, I dare say, when they didn't happen to cross each other's wishes. No, it's the implanted goodness that saves,—the seed of right-eousness treasured from generation to generation, and carefully watched and tended by disciplined fathers and mothers in the hearts where they have dropped it. The flower of this implanted goodness is what we call civilization, the condition of general uprightness that Halleck declared he owed no allegiance to. But he was better than his word." [pp. 332-33]

The doctrine that Atherton expounds blends elements from two intellectual traditions: Horace Bushnell's doctrine of "Christian nurture," and the secular doctrine of progress that was an integral part of the dominant American ideology. Bushnell had asserted that the operation of divine grace need not take the form of a spectacular conversion occurring at a specific time and place, but might be transmitted through a line of pious ancestors by whose merits "the seed of a regenerate life is implanted" in the child at birth.[45] This is an essentially patrician theory, developed in opposition to the revival movement that was sweeping the country during the 1840s when Bushnell first published his doctrine. It offers supernatural support for a hereditary hierarchy of social classes. Furthermore, it fosters what might be called a bureaucratic conception of the distribution of divine favor through established channels. Atherton's term "civi-lization" refers to the notion that every human society evolves through a fixed succession of social stages from the primitive bar-barism of prehistoric man to the highest level of culture.[46] This theory implied that at any given moment the classes of any highly developed society exhibit the various stages of social evolution, from the barbarity of the lowest class to the refinement of the highest. Thus the idea of civilization also supported an aristocratic class-con-sciousness.

Atherton's explanation of the difference between the Hubbards (especially Bartley) and Ben Halleck is that Halleck has back of him several generations of righteous and disciplined ancestors. But the ancestors, if we look back of Ben's parents, are entirely hypothetical; they receive no other mention in the novel. And Atherton's ideology fails to match Halleck's case in an even more important respect. For Ben's own account of his salvation makes no reference to social obligation, but rests on a fundamentalist conception of what seems to be emphatically an amazing grace:

Passage J
In entering the ministry [says the narrative voice] he had returned to the faith which had been taught him almost before he could speak. . . . He freely granted that he had not reasoned back to his old faith; he had fled to it as to a city of refuge. His unbelief had been helped, and he no longer suffered himself to doubt; he did not ask if the truth was here or there any more; he only knew that he could not find it for himself, and he rested in his inherited belief. He accepted everything; if he took one jot or tittle away from the Book, the curse of doubt was on him. He had known the terrors of the law, and he preached them to his people; he had known the Divine mercy, and he also preached that. [p. 359]

There is a good deal of represented discourse here, but the effect is drastically different from that of Passage A (above). The appropriately high incidence of biblical language helps to establish the tone of the prose. The effect, however, is due mainly to the fact that instead of maintaining an ironic distance, as Howells does in Passage A, here he is closely identified with the thoughts and feelings he describes, to such an extent that no irony is directed toward the sonorous unction recalling the pulpit oratory of an earlier day. For the theology of the passage is virtually identical with that of a conservative pre-Civil War evangelical Christianity, quite unlike the liberal Protestantism of Beecher.

The last scene of *A Modern Instance* takes place a year after word of Bartley Hubbard's death has reached Boston. Atherton and his wife in their summer place "on the Beverley shore" (p. 360) are discussing a letter from Halleck in which he asks Atherton to advise him whether, after this lapse of time, he does not have the moral right to ask Marcia to be his wife. When Clara Atherton urges her husband to send his approval, he hesitates, saying: "There might be redemption for another sort of man in such a marriage; but for Halleck there could only be loss,—deterioration,—lapse from the ideal" (p. 362). And when his wife presses him he can only sigh, "Ah, I don't know! I don't know!" This has been praised as an "open" ending, and it does break away from the Victorian convention of tying up all loose ends of plot in the last chapter of a novel. It is true also that Atherton's uncertainty makes him seem a little less of a prig than he has seemed earlier. But his answer to the question is less important than the fact that Howells gives the matter so much prominence, as if it were one of the major problems generated by urban industrialism in this country in the later nineteenth century. Neither Atherton's genteel rationalism nor Halleck's desperate retreat into religious fundamentalism leaves an impression of profound insight on their part, or on Howells's.

At the end of the 1880s, Howells made an even more ambitious effort to test his theory that honest portrayal of carefully observed conditions, even in the apparent chaos of contemporary society, would reveal a system of moral principles controlling events. This effort, *A Hazard of New Fortunes* (1890), is much larger in scope than is *A Modern Instance,* and Howells had realized in the intervening years that the moral problems generated by economic and social forces were of greater human consequence than the genteel question of whether unspoken love for a married woman is adulterous. In the later novel, laid in contemporary New York City, a young Christian Socialist and a German radical who has chosen to devote himself to the poor are killed in a street riot growing out of a strike of horsecar workers. But their deaths are in effect accidental and therefore devoid of economic or social meaning. At the end of a long and often brilliantly written book, Basil March, serving once again as a spokesman for the writer, pronounces judgment to his wife: the German radical has been punished for advocating violence, but the Christian Socialist died because "it was his business to suffer there for the sins of others." "Isabel," continues Basil, "we can't throw aside that old doctrine of the Atonement yet. The life of Christ, it wasn't only in healing the sick and going about to do good; it was suffering for the sins of others! . . . If we love mankind, pity them, we even *wish* to suffer for them."[47]

This may be good theology—although I doubt it—but it implies a passivity, even a kind of masochism, that allows no place for a serious criticism of society or its established system of values. During the thirty years of Howells's writing career that remained to him, he made a few gestures toward challenging prevalent assumptions: the utopian romances about Altruria, certain abrasive touches of social satire in *The Landlord at Lion's Head* (1898), a commentary on the gullibility of the followers of a self-proclaimed religious leader in *The Leatherwood God* (1916). But he did not find a way of shattering the middlebrow mold from which he had set out to free the American novel. Two of Howells's contemporaries and lifelong friends, Mark Twain and Henry James, in their very different ways, came nearer. And it was they, not he, who transmitted to American novelists of the twentieth century what the next generation could use of the achievements of their predecessors in the nineteenth century.

Guilt and Innocence in Mark Twain's Later Fiction

Like other American businesses, publishing became organized on a national scale during the decades following the Civil War. The varied components of the reading public began to be recognized and the mechanisms for the production and distribution of the staple commodity, fiction, were adjusted more precisely to the demands of the market. Howells's audience, for example, can be identified with some confidence: he was a central figure in the world of letters defined by the subscription lists of the major monthly literary magazines that dominated the field of publishing from the 1870s to the 1890s. He was hired as assistant to James T. Fields on the editorial staff of the *Atlantic Monthly* in 1866, was made editor-in-chief in 1871, and continued in that position for a decade, until he was well enough established to support his family by free-lancing as novelist and critic with only part-time editorial work. During the early 1880s he contributed frequently to the *Century*, but in 1885 he signed a contract to write regularly for *Harper's Monthly*, and was associated with that journal in various capacities down to his death in 1920—although from time to time he took on editorial functions for other magazines.[1]

The big monthlies—especially *Atlantic*, *Century*, *Scribner's*, and *Harper's*—had a combined circulation that ranged, with much overlap, from perhaps 150,000 in the 1870s to twice that figure in

1890;[2] thereafter their circulation declined in the face of competition from cheaper and less genteel magazines such as *Cosmopolitan, Collier's, McClure's,* and *Munsey's.*[3] Down to 1891, when a copyright agreement between the United States and Great Britain was at last negotiated, the sale of novels by native writers in book form was hampered by the competition of cheap pirated reprints of English novels, which American publishers could bring out without paying any royalties. American writers were obliged to depend heavily on returns from serial publication in magazines. Thus after Howells's *A Hazard of New Fortunes* (1889) had been serialized in *Harper's Weekly* (with a circulation of some 200,000), he was delighted by a sale of only 18,000 for the novel during the first year.[4]

Serial publication provided a relatively stable market, but it imposed significant constraints on novelists. Even though the subscribers to the monthly magazines were less naïve than the readers of the New York *Ledger* (which continued to thrive during the 1870s and 1880s), they nevertheless represented the middlebrow system of values that George Santayana would later identify as the genteel tradition. The Norwegian-born immigrant writer Hjalmar Boyesen protested in 1887 that the refusal of the magazines to publish any word not suitable for reading aloud around the family fireplace was stifling American fiction.[5] An even more damaging restriction was the pressure exerted on novelists to avoid any material that was intellectually challenging.[6]

Mark Twain had some contact with the literary magazines. Howells published several specimens of his friend's work in the *Atlantic* during the 1870s (for example, the seven installments of "Old Times on the Mississippi"), and Mark Twain declared he liked to write for the *Atlantic* audience because it did not "require a 'humorist' to paint himself stripèd & stand on his head every fifteen minutes."[7] Furthermore, in the 1880s Richard Watson Gilder ran in the *Century* three installments of selections from *Adventures of Huckleberry Finn* (carefully edited to remove unsuitable expressions).[8] But Mark Twain was not really at home in this genteel atmosphere. His uneasiness in such company had been dramatically revealed in the humorous after-dinner speech he delivered at a banquet given by the publisher of the *Atlantic* in honor of the seventieth birthday of John Greenleaf Whittier in 1877. The little talk—which suggested that both Clemens himself and certain venerable New England Men of Letters (i.e., Emerson, Longfellow, and Holmes) might be imposters—was enjoyed by highbrows such as Francis J. Child of Harvard but outraged middlebrows in inland Massachusetts and the Middle West. Significantly, it seemed to Howells a "hideous mistake" because of its irreverence.[9]

This incident showed that Mark Twain had been intuitively right at the outset of his career when he had chosen to address himself to an audience quite different from the readers of the literary monthlies. In 1867, soon after he had returned from the *Quaker City* excursion to Europe and the Holy Land, he had received from Elisha Bliss, president of the American Publishing Company of Hartford, a letter proposing that he make a book from the dispatches he had sent back to newspapers.[10] The American Publishing Company was what was known as a subscription firm. That is, its publications were sold by house-to-house canvassers rather than through the regular channels of the book trade. The plan had two great commercial advantages: it eliminated the retail bookseller's profit, and it reached the vast majority of the population that never entered a bookstore and perhaps did not live within geographical reach of one. Down to the entry of Mark Twain into the field, subscription books had been entirely non-fictional: religious works, household medical encyclopedias, first-person narratives of travel or adventure such as David Livingstone's journals of his explorations in Africa, and autobiographies of celebrities such as P. T. Barnum.[11] The subscription audience was fond of pictures and not discriminating in its appreciation of them.

The Innocents Abroad, the book based on Mark Twain's letters about the *Quaker City* excursion, was published in 1869 and was an instant success. Bliss sold 78,000 copies of it in sixteen months, 100,000 in three years, and another 30,000 copies during the next seven years. Between 1869 and 1881, when Bliss died and Mark Twain ended his connection with the firm, the American Publishing Company paid the writer $105,000, representing sales of more than 300,000 copies of his books.[12] This was a large sum of money a century ago. On the other hand, Mark Twain's commitment to subscription publishing had serious disadvantages for him. It typed him as a writer for a subliterary audience and may have led some sophisticated readers to avoid his work altogether. The attitudes of the subscription audience probably encouraged his undeniable lapses in taste and occasional careless writing. He told his friend Joel Chandler Harris: "Mighty few books that come strictly under the head of *literature* will sell by subscription."[13]

Superficially, then, it may be said that Mark Twain set out to make money by writing for a lowbrow audience, an audience that had no interest in literature as that term had been understood. Yet this was probably a wise decision. It was a way for the writer to free himself from the dominant literary conventions. In doing so Mark Twain was following the example of all his truly first-rate predecessors—Emerson, Hawthorne, Melville, Thoreau, Whitman. Like these other pioneers, however, he exposed himself to grave dangers. Lacking the support of a technical literary tradition, he was forced to invent a new form and a new style in which to express himself, or at least to make the drastic revisions in existing genres that were necessary to adapt them to his purposes. And in the absence of the support a writer has the right to expect from perceptive criticism, Mark Twain could hardly be expected to arrive at a clear grasp of the difficulties he faced. For example, he needed to understand the impasse exhibited in Beecher's *Norwood:* to recognize that Hiram Beers testing the speed of the horses left outside the meeting house during the service embodies more personal warmth, represents a closer approach to the fullness of human potentiality, than do the shadowy churchgoers dozing through the sermon inside the church. This intellectual feat was prerequisite to the reconstruction of the form of the novel. A structure had to be devised in which Beers, in his capacity as representative of the low characters, could be given a central position and the high or "grammatical" characters could be relegated to the margins or dropped altogether. And of course the motive power for such a change (as Howells's term "grammatical characters" implies) would have to be provided by a revolution in the language of fiction.

This is what Ernest Hemingway meant by his often-quoted statement that "all modern American literature comes from one book by Mark Twain called *Huckleberry Finn*,"[14] and what George Santayana was saying from a different perspective when he bracketed the humorists with Whitman as writers who had escaped, at least partially, from the genteel tradition.[15] Yet both these critics qualified their remarks: Hemingway added that one must stop reading *Huckleberry Finn* at the point where Tom Sawyer appears on the Phelps plantation, and Santayana said that the humorists could not entirely escape the genteel tradition because they had nothing solid to put in its place. This is to say that even Mark Twain's best work contains significant vestiges of the outworn system of values that I have called middlebrow. I propose to test these judgments by undertaking something like a biopsy, a microscopic examination of a celebrated passage from the novel in order to determine if possible what traces of infection from the popular culture can be found there.

In a study of Mark Twain published some years ago,[16] I argued that his revolutionary accomplishment could best be understood as the representation of increasingly complex modes of experience from a "vernacular perspective." His point of departure as a writer had been the work of humorists of the previous generation such as Thomas B. Thorpe, author of "The Big Bear of Arkansas," and George W. Harris, author of the Sut Lovingood tales, who had set about reproducing in print the oral effects of backwoods storytellers. This native humor typically involves two main characters: an uncultivated rustic, and a "straight" or (relatively) cultivated character who provides a framework of introduction and conclusion for the yarn or tall tale of the vernacular character. The straight character speaks in correct, even pedantic or pompous language which contrasts vividly with the incorrect but highly colored speech of the backwoods character. In Mark Twain's best writing (including of course *Adventures of Huckleberry Finn*) the vernacular spokesman takes over the narrative entirely; the straight character disappears and although his presence can still be felt behind the scenes or beneath the surface, the speech of the vernacular character becomes the only available narrative medium. But even when the straight character is no longer visible, there is an implied antagonism between him and the vernacular character. The antagonism represents a polarity between the high culture of the society, and a folk culture based on a contrasting system of values. A polarity of this sort seems to be a structural feature of American culture as a whole during the later nineteenth century.

Santayana, for example, in the discussion of the American humorists I have referred to earlier, was concerned with "indications of a truly native philosophy and poetry" which may have arisen in this country "to express . . . the spirit and the inarticulate principles that animate the community, on which its own genteel mentality seems to sit rather lightly."[17] This observation follows a historical parallel that I think has not received the attention it deserves:

> Perhaps the prevalence of humor in America, in and out of season, may be taken as one more evidence that the genteel tradition is present pervasively, but everywhere weak. Similarly in Italy, during the Renaissance, the Catholic tradition could not be banished from the intellect, since there was nothing articulate to take its place; yet its hold on the heart was singularly relaxed. The consequence was that humorists could regale themselves with the foibles of monks and of cardinals, with the credulity of fools, and the bogus miracles of the saints; not intending to deny the theory of the church, but caring for it so little at heart, that they could find it infinitely amusing that it should be contradicted in men's lives, and that no harm should come of it. So when Mark Twain says, "I was born of poor but dishonest parents," the humor depends on the parody of the genteel Anglo-Saxon convention that it is disreputable to be poor; but to hint at the hollowness of it would not be amusing if it did not remain at bottom one's habitual conviction.[18]

It will be observed that Santayana considers the "truly native philosophy and poetry" of America to be more valuable than the traditional philosophy and poetry, and that he sees a conflict between them which is still undecided but in which the native philosophy and poetry are bound to prevail.

When Santayana speaks of escaping from the genteel tradition, he is referring to a kind of transcendence. (Perhaps it might be called, paradoxically, a transcendence downward.) Let me illustrate the process in Mark Twain's work by examining not an overtly comic passage but a comparatively serious page or so from the celebrated Chapter 31 of *Adventures of Huckleberry Finn*. In this chapter, it will be recalled, Huck discovers that the rascally Duke and King have turned in his friend Jim as a runaway slave, collecting part of the reward offered in a fraudulent handbill concocted by the Duke. The necessity of taking some action forces Huck for the first time to face up to the fact that his help to this fugitive defies the stern imperatives of the culture in which he has grown up. His conscience, he says, "went to grinding" him, and made him feel "lowdown and ornery."

> And at last, when it hit me all of a sudden that here was the plain hand of Providence slapping me in the face and letting me know my wickedness was being watched all the time from up there in heaven, whilst I was stealing a poor old woman's nigger that hadn't ever done me no harm, and now was showing me there's One that's always on the lookout, and ain't agoing to allow no such miserable doings to go only just so fur and no further, I most dropped in my tracks I was so scared.[19]

This passage contains a number of vernacular phrases and images, such as "most dropped in my tracks" and "low-down and ornery." The confusion about the antecedents of relative pronouns (suggesting, for example, that the nigger hadn't ever done Huck any harm and was showing him the vigilance of Providence); the double negatives; and the omission of a subordinating conjunction between the last two clauses in the sentence are also unmistakably vernacular in flavor. The most brilliant effect of the passage, however, depends on the discourse pronounced by Huck's conscience that is not quoted directly but is given us only in Huck's paraphrase. It is a complex example of represented discourse. The theological ideas reported by Huck are clearly not his own, and in the unctuous periphrasis of the reference to "One that's always on the lookout" even the very diction of the minatory conscience glitters incongruously through the texture of Huck's vernacular speech. In this fashion the notion that slavery is sanctioned by Christianity is linked with the dominant culture of the pre-Civil War South, and the language in which it is stated shows that culture to be debased and at the same time comically pretentious. The language Huck uses demonstrates that he does not really understand the theological justification of slavery, but he accepts what he understands of it without reservations. Conscience, which Mark Twain's audience in the 1880s believed to be the voice of God or an infallible Moral Sense, is revealed as merely the internalized mores of the community. And since the community in question, that of Huck Finn's fictive world, is corrupted by slavery, a conscience functioning as its spokesman becomes the voice of evil rather than a divine monitor.

The moral contamination to which Huck has been exposed is depicted with incomparable skill in his own words. Of course, throughout the story the reader is presumed to accept the convention of first-person narrative which allows the narrator exact and total recall of his own remarks and those of all other characters. When the narrator functions in this way, he becomes a merely mechanical, morally neutral device for transmitting to the reader what was said. But Huck is not quoting the supposed speech of his conscience; he makes it partly his own by paraphrasing it. The speech is composed of remembered fragments of the teachings of Miss Watson and other instructors in morals. When Huck paraphrases it he adopts the bombastic cadences as part of his own style, although in the process they are inevitably colored by his pronunciation and syntax. This distortion of Huck's language registers the invasion of his mind by the environmental force of the dominant value system and poses a correspondingly greater barrier to his attainment of the ideal condition of inner freedom which (in the novel's ethical system) is synonymous with true happiness.

Long after *Adventures of Huckleberry Finn* was published, Mark Twain called it "a book of mine where a sound heart & a deformed conscience come into collision & conscience suffers defeat."[20] So far we have been concerned with the portrayal of the deformed conscience. After this phase of Huck's inner debate, when he has decided to rid himself of his burden of sin by notifying Miss Watson where she can find her runaway slave, the sound heart is aroused. Huck's consciousness is flooded with emotionally charged memories of the time he has spent with Jim on the raft, particularly memories of Jim's kindnesses to him:

> . . . I see Jim before me, all the time, in the day, and in the nighttime, sometimes moonlight, sometimes storms, and we a floating along, talking, and singing, and laughing. . . . I'd see him standing my watch on top of his'n, stead of calling me, so I could go on sleeping; and see him how glad he was when I come back out of the fog. . . . [p. 271]

The passage is too long to quote in full, but it will be familiar to most readers as one of the high points in our literature.

Nevertheless, a more searching analysis of Chapter 31 shows that Mark Twain's dazzling effect has been achieved only at a certain cost. As Alan Trachtenberg has observed, the writer faced a difficult technical problem in reconciling two different roles that Huck is obliged to play. Huck is both "the verbalizer of the narrative" and "a character within the narrative."

> Do Huck's traits [asks Trachtenberg] derive in fact from an inner life at odds with social necessity, or from . . . imperatives of his role as narrator? Obviously we need not make an either/or choice. But the second alternative has been so little present in criticism it is worth considering at some length. The crux of the matter is whether . . . a sentient inner life is always present.[21]

Up to Chapter 31 Huck's developing consciousness resembles that of Donatello in *The Marble Faun:* like Donatello he seems to be undergoing a transition from childlike innocence to the moral depth of maturity in consequence of a growing recognition of the conflict between his responsibility toward Jim and the demands of society. Like Donatello, he is being educated by guilt. This education reaches its apex in the dialogue with his conscience. But Huck's situation is much more complicated than that of Donatello. Whereas Hawthorne allows no doubt concerning Donatello's guilt (he has entered into a sexual relation with Miriam that Hawthorne regards as tainted, and as a symbolically inevitable consequence has committed murder), Mark Twain intends to call in question the validity of the moral and legal system that supports slavery; Huck's guilt is ultimately not guilt at all, but only a delusion, an aspect of his profound innocence. More interestingly, perhaps Huck himself does not take the burden of guilt he is assuming with complete seriousness. In the first chapter of the book, when Miss Watson begins her religious instruction of Huck, he says:

> Then she told me all about the bad place, and I said I wished I was there. She got mad, then, but I didn't mean no harm. All I wanted was to go somewheres; all I wanted was a change, I warn't particular. She said it was wicked to say what I said; said she wouldn't say it for the

> whole world; *she* was going to live so as to go to the good place. Well,
> I couldn't see no advantage in going where she was going, so I made
> up my mind I wouldn't try for it. [p. 19]

Furthermore, whatever Huck may believe, the reader whom Mark
Twain was addressing in the 1880s must be assumed to have regarded the wickedness of slavery as an axiom. Both author and
reader understand very well that Huck has a sound heart; it is his
conscience that is depraved. Within this frame of assumptions,
Huck's decision to help Jim is so unassailably virtuous that his
supposed wickedness can hardly fail to seem comic.

This aspect of the situation tends to call into play Huck's role as
a dead-pan vernacular narrator, whose function is to exploit comic
possibilities rather than to explore moral issues. Accordingly, in the
original version of the scene (as Walter Blair has pointed out), Huck
expanded his resolution to go to hell in the manner of a tall tale:

> I would take up wickedness again, which was in my line, being brung
> up to it. What I had been getting ready for, and longing for and pining
> for; always, day and night and Sundays, was a career of crime. And
> just that thing was the thing I was a-starting in on, now, for good
> and all.[22]

These words, which appear in the surviving manuscript of Chapter
31, were wisely deleted in revision, but they throw light on Mark
Twain's attitude toward Huck's guilt.

The solemn, biblical simplicity of "just that thing was the
thing I was a-starting in on" is a reminder that the traditional cult of
oratory continued to flourish in Huck's South long after it had
begun to wane in other regions. When he describes his effort to pray
for forgiveness (in itself a rather self-consciously theatrical gesture)
his syntax takes on the rhythms of pulpit and rostrum:

> . . . I kneeled down. But the words wouldn't come. Why wouldn't
> they? It warn't no use to try and hide it from Him. Nor from *me*,
> neither. I knowed very well why they wouldn't come. It was because
> my heart warn't right; it was because I warn't square; it was because
> I was playing double. I was letting *on* to give up sin, but away inside
> of me I was holding on to the biggest one of all. [p. 270]

The boy who utters these pompous sentences is the legitimate heir of the Pap whom the "new judge" in St. Petersburg had set about reclaiming from his drunkenness:

> . . . when it was bedtime, the old man rose up and held out his hand, and says:
> "Look at it gentlemen, and ladies all [no one is present but the judge and his wife]; take ahold of it; shake it. There's a hand that was the hand of a hog; but it ain't so no more; it's the hand of a man that's started in on a new life, and 'll die before he'll go back. You mark them words—don't forget I said them." [p. 43]

In analyzing Huck's debate with his conscience we must take into account yet other feelings that in the 1880s were inevitably called into play for Americans by the topic of slavery. These were related to the complacency that had come to permeate public discourse in the North during the fifteen or twenty years since Appomattox—years during which the articulate spokesmen for the society had developed retroactively a set of war aims and a conception of how the war had been fought, an ideology designed—consciously or unconsciously—to console the victorious section for the half-million casualties it had suffered in the war and for the conspicuous collapse of high-minded Reconstruction policies during the subsequent decade. The emotions clustering around these topics were so strong that an extraordinary effort would have been required to resist their contagion. And Samuel Clemens was not disposed to resist. He surrendered uncritically to the mood of the New England society in which he had chosen to live. Although in the cynical "compromise of 1877" the leaders of the Republican party had in effect allowed the South to resume control over the freed slaves, these same party leaders had carefully kept alive the passions aroused by the War and the belief that it had been primarily a crusade to end slavery. In the ritual known as "waving the bloody shirt," Republican political orators had continued year after year to dwell upon memories of Union casualties as a means of attracting voters away from the Democratic party—which was just as regularly re-indicted for the Southern crimes of slavery and disunion. Waving the bloody shirt, if it had ever been a legitimate maneuver, had become with the passage of years a systematic self-deception on a vast scale. Huck's decision to help Jim was bound to be viewed by both Mark Twain and his readers through this lens. It was part of the moral degradation that every war leaves as its legacy—and like all such degradation, it generated an intense self-righteousness.

In November 1879, at the halfway point in the long gestation of *Adventures of Huckleberry Finn*, Mark Twain was invited to speak at a "Grand Re-Union of the Army of the Tennessee" in Chicago. His own brief remarks, which ventured upon a daring although in the end laudatory joke at the expense of General Grant, were a triumphant success, but that is beside the point here. I am interested rather in Clemens's emotional response to the unabashed flag-waving. Upon his return home he wrote to his friend Howells a remarkable letter describing the banquet:

> I doubt if America has ever seen anything quite equal to it; I am well satisfied I shall not live to see its equal again. . . . Bob Ingersoll's speech was sadly crippled by the proof-readers [Clemens enclosed a printed copy], but its music will sing through my memory always as the divinest that ever enchanted my ears. And I shall always see him as he stood that night on a dinner table, under the flash of lights & banners, in the midst of seven hundred frantic shouters, the most beautiful human creature that ever lived.

Like other speakers, Ingersoll engaged in a simplification and revision of the extremely various and complex motivations of the Union armies. And it should be remembered also that he was speaking fourteen years after the end of hostilities—as if a celebrated orator had been evoking before a convention of the American Legion in 1932 the slogan of making the world safe for democracy. Mark Twain continued with a remembered quotation from Ingersoll's speech:

> "They fought that a mother might own her own child"—the words look like any other print, but Lord bless me, he borrowed the very accent of the angel of Mercy to say them in, & you should have heard the hurricane that followed.[23]

In the printed version of Ingersoll's speech the passage reads as follows:

Grander than the Greek, nobler than the Roman, the soldiers of the Republic, with patriotism as shoreless as the air, battled for the rights of others, for the nobility of labor, fought that mothers might own their babes, that arrogant idleness might not scar the back of patient toil, and that our country should not be a many-headed monster made of warring states, but a nation, sovereign, great, and free.[24]

Huckleberry Finn's conclusion—"All right then, I'll *go* to hell"—is tainted by this kind of hyperbole. I have suggested that in the triumph of Huck's vernacular language in this scene—and it is an undeniable triumph—close inspection can nevertheless detect a touch of the self-delusion and complacency that Mark Twain took such contagious delight in satirizing in Pap Finn, or the Duke and the King. There is a trace of false feeling in Huck's response to his conscience, related (I think) to Mark Twain's inability to develop fully the tragic potentialities of Huck's situation. To put this observation another way: there is a covert suggestion that Huck is dramatizing himself, that he has an inkling of the falsity of the moral stance of his conscience. (After all, it was Huck who had called the King's discourse before the coffin of Peter Wilks, the tanner, "soul-butter and hogwash.")

In any case, at the end of Huck's debate with his conscience, Mark Twain faces a formidable dilemma in the conduct of the narrative. If Huck is primarily a dead-pan comic mask, he cannot be made into a tragic protagonist because he has no inner life of any consequence. If he is a completely innocent hero confronting absolute evil, as in a melodrama, he is equally ineligible for a tragic role, because his inner life is too simple. And whatever Huck may think or feel, there is no plausible way for Jim to be rescued. Mark Twain responds to this crisis in much the same way he had responded to a similar impasse much earlier, when Huck and Jim found out they had drifted in the night past Cairo at the mouth of the Ohio. He creates a totally new narrative situation. He dispenses with the Duke and the King (they will be seen a few pages later being ridden out of town tarred and feathered on a rail) and reintroduces Tom Sawyer, who has been absent from the story (except of course in Huck's thoughts) since Chapter 3. We shall eventually be asked to believe that Tom brings with him the knowledge that Jim has been set free by Miss Watson on her deathbed: the problem that has provided the focus of the plot through most of the book is wiped out in a moment, and the story of Jim's escape is brought to an end as if it were of only peripheral concern. But for the time being, Tom keeps his information to himself in order to carry out his own project, his version of how the escape of a slave (that is, a prisoner) ought to be managed. This version can blossom into whatever absurdities occur to him because the human and moral significance of Jim's escape has been eliminated. The reintroduction of Tom into the story places it within a framework of assumptions proper to the popular culture of the postwar North rather than to the prewar South because slavery is no longer a problem. On its much smaller scale, Miss Watson's magical change of heart is a functional equivalent in the plot to the historical outcome of the War: both events are assumed to have solved all problems. Since the Evasion that Tom plans and executes is stripped of practical relevance, it can be simply entertainment, an "effect," a "happening" that foreshadows the "conceptual art" of the 1960s. It is literally absurd. Perhaps this removal of the theme of slavery and escape from the level of practical behavior to that of esthetic spectacle was already implicit in the rhetorical exaggeration and the over-acting which, with the benefit of hindsight, we can identify in the account of Huck's debate with his conscience.

The drastic shuffling of identities at this point in the book has often been noticed. Although Huck Finn continues to tell the story in the first person, the moral atmosphere is established by Tom because Huck surrenders his own plan for rescuing Jim as well as his own criteria of judgment, submitting himself obediently to Tom's instructions with only an occasional murmur of skepticism. Mark Twain registers the change by contriving to have Huck bear Tom's name. As a consequence, the theme of guilt disappears and Huck's debate with his conscience is forgotten; there is no place for such matters in Tom Sawyer's universe. No longer a runaway slave, Jim becomes an imprisoned aristocrat analogous to the Man in the Iron Mask or the Prisoner of Chillon. And a scrupulous observance of the supposed "rules" for delivering such aristocrats from prison (that is, an empty formalism) takes the place of moral principles.

Tom Sawyer's fantasy world is permeated by the spurious emotions of cloak-and-sword fiction, evoked by the same language. This
is demonstrated in the scene where, with an irresistible overflow of
comic inventiveness, Mark Twain exhibits Tom being taken in by
the false pathos of the "mournful" inscriptions he himself has
composed for Jim to carve upon the walls of his cabin: "Here a
captive heart busted"; "Here a poor prisoner, forsook by the world
and friends, fretted out his sorrowful life"; and several others.
"Tom's voice trembled, whilst he was reading," says Huck, "and he
most broke down" (pp. 325–326). The ten chapters devoted to Tom
Sawyer's Evasion can be thought of as a comic version of the theme
treated seriously in the account of Huck's defiance of his conscience.
The effect is somewhat like that of an Elizabethan double plot, or
more precisely, the gravediggers' scene in *Hamlet*. In the satiric
middle section of the book, a case against the culture that Mark
Twain ascribes to the prewar South has been gradually built up
through the depiction of Pap Finn's taking the pledge in the house
of the new judge, or Miss Watson's lessons in theology (Chap. 3), or
the Grangerford-Shepherdson feud (Chap. 18), or Emmeline Grangerford's works of funerary art (Chap. 17),[25] or the King's address to
the mourners before the coffin of Peter Wilks (Chap. 27), or—
climactically—the homily about slavery delivered by Huck's conscience. These various aspects of the high culture of the region—
fragmented, outworn, deprived of all meaning except their availa-
bility for constructing rationalizations—have in common a bookish
quality. In *Life on the Mississippi*, completed during an interval in
Mark Twain's work on *Adventures of Huckleberry Finn*, he had
advanced the explicit thesis that the culture of the South had been
poisoned by Walter Scott's Waverley novels (Chap. 46), and Tom
Sawyer's favorite melodramatic romances are vulgarizations of Scott.
Thus the "rules" of procedure controlling Tom's fantasy world,
derived from literary tradition rather than from experience, have a
structural position and function resembling that of the system of
theology and law which in Huck's view had condemned him to hell.
It now appears that Huck's transcendence of this dominant
value system has been implicit from the outset in Mark Twain's
choice of Huck's vernacular speech as the narrative medium for the
entire story. Academically correct speech, especially if it is in the

slightest degree exalted, has been systematically linked in this book with hypocrisy, self-dramatization, fraudulent claims to status, cynicism, and cruelty, all radiating outward (so to speak) from the institution of slavery. Huck's perspective escapes the control of the dominant popular culture by virtue of the fact that his vocabulary lacks the high-sounding abstract terms (such as property, sin, and Providence) in which it is incorporated and without which its values cease to exist. Thus his vernacular speech not merely expresses a state of mind characterized by inner freedom, it embodies that state of mind, indeed it *is* that state of mind. The primary achievement of *Adventures of Huckleberry Finn* is its language; through Huck Finn, Mark Twain made available to the next generation of writers a literary prose freed from the associations and connotations of the decadent high culture of the nineteenth century. This prose (and the cultural revolution it embodies) is what the successors of the native humorists, if not the humorists themselves, had to put in place of the genteel tradition.

Mark Twain's magnificent stylistic achievement, however, has no direct bearing on slavery or the larger issue of the corruption of the culture of the Old South. The comic conclusion of the book leaves unresolved the issues raised by his criticism of the value system focused on slavery, and the several sequels in which the writer undertook to recount further adventures of Huck, Tom, and Jim demonstrate that the escape into fantasy represented by Tom's Evasion was a completely sterile mode of transcendence. For his next book Mark Twain turned instead to a theme he had discovered in writing the second part of *Life on the Mississippi.* As he moved up the River from St. Louis, he had been deeply impressed by the contrast between the Southern towns along the banks and the brisk, businesslike attitudes, the general air of prosperity, the neatness and fresh paint characteristic of the towns along the River in Iowa, Illinois, and Wisconsin. In a minutely analytical study of *Life on the Mississippi,* Horst H. Kruse has identified the contrast between Southern backwardness and Northern enlightenment and prosperity as a basic structure. The relation of South to North became in the mind of the writer the relation of past to present, and this relation exemplified for him the doctrine of material progress which in turn engendered moral progress, a historical movement toward the ideal of civilization. Indeed, Mark Twain began to view himself as performing a function like that performed by Cervantes in *Don Quixote* —the destruction of medieval romance with its reactionary cult of chivalry and aristocracy, by means of satire.[26] We recognize here again, of course, yet another evidence of Mark Twain's adoption of attitudes current in the complacent popular culture of the victorious North. With Cervantes in mind, we can perhaps see a further meaning in the final chapters of *Adventures of Huckleberry Finn*: the protagonist is acting out a literary burlesque of historical romances. But the topic of literary romanticism seems trivial as a target in comparison with the grim theme of slavery that it supplants in the final sequence.

Mark Twain himself seems to have been vaguely aware that in *Adventures of Huckleberry Finn* he had posed a problem of vast implications but had left it dangling in the air. Almost coincidentally with the publication of that book he recorded in his notebook a dream that became the germ of his next work. Like Tom Sawyer's Evasion, this was also to be a literary burlesque; and it returned to the theme that Mark Twain described as "a contrast of civilizations."[27] The new work would eventually appear as *A Connecticut Yankee in King Arthur's Court* (1889). It was an imaginative probing of the implications of the idea of progress. Because I have discussed this book also at length elsewhere, I shall observe here only that introducing a vernacular protagonist to a corrupt society—one resembling in many ways the Old South depicted in *Adventures of Huckleberry Finn*—is a storyteller's way of raising once again, on a much larger scale, the question Mark Twain had confronted and then had turned away from in the previous book. He could not find a way for Huck (and Jim) to become integrated with the corrupt society they had observed in the towns along the River.[28] But these towns had been part of an archaic social order. Might not the case be entirely different if the society in question was modern, advanced, progressive? If there was no possibility for a barely literate orphan boy to reform and regenerate these backward towns of the Old South, it was nevertheless evident that some force had been able to give to the towns of the northern part of the Mississippi Valley a totally different social system.[29] In the North, history had proved capable of transforming society, bringing enlightenment and democracy. It seemed to Clemens that the crucial difference between backward South and progressive, democratic North was progress (in the first instance economic, but inevitably also political and moral).

At any rate, this was the simplified form in which Mark Twain's imagination was able to represent a cluster of ideas making up the dominant secular faith of the American North in the later nineteenth century. It was widely believed that the appointed way for mankind to escape the poverty and injustice of its medieval past was the advance of technology, bringing about an industrial revolution as exemplified in the society of the triumphant Union. Furthermore, the theme of technological advance could readily be merged with the vernacular perspective by presenting both of them as forces hostile to the decadent high culture of the nineteenth century, which in turn could be portrayed as a long-established cultural tradition surviving anachronistically into the present. In this pattern, progress implied a rejection of high or genteel culture: Hank Morgan, the Promethean Yankee from Connecticut, was to be a self-proclaimed philistine. "I am a Yankee of the Yankees," he announces, "and practical; yes, and nearly barren of sentiment, I suppose—or poetry, in other words."[30] His immunity from the corruptions of Arthurian Britain is, so to speak, built in: he is a transcendent figure by virtue of the fact that he represents a later century, separated from the dark ages by more than a thousand years of technological, moral, social, and political development.

In an earlier essay I described *A Connecticut Yankee* as "Mark Twain's Fable of Progress." The "contrast of civilizations" that he intended to develop was axiomatically favorable to the nineteenth century in comparison with all eras of the history of mankind before the American and French revolutions in politics, and the industrial revolution in economics. In the person of Hank Morgan, archetypal Yankee, the United States stands for the present as against the past, which is represented by a Britain whose monarchy and aristocracy perpetuate medieval institutions and customs. Dan Beard, the artist whom Mark Twain chose to illustrate the first edition of the book and whose work he ecstatically approved, clothed the Yankee in the striped trousers and top hat of Uncle Sam. This iconography is richly suggestive. It reminds us that in the self-confident American nationalism of the nineteenth century, the turning-point in human history was asserted to have come with the formation of the Republic. It remained only for the other nations of the world to follow this example.

Mark Twain's fable depicts an American Prometheus creating a new race of men by bringing to them the fire of technology and the light of political justice and democracy. It is true that the writer's economic naïveté led him to ignore the question of how the Yankee's industrial revolution was to be financed, but the intention is grandiose. As the story develops, the literary burlesque of the *Morte d'Arthur* becomes a kind of Capitalist Manifesto, promising a revolution that will eliminate poverty and oppression, and through the discipline of technology, effect the moral and intellectual regeneration of the entire populace. Like Tom Sawyer's Evasion, that other literary burlesque, this story belongs unmistakably to the genre of romantic fantasy: there is no place in it for novelistic characters. After some hints in the early chapters that Hank Morgan is merely a 'cute Yankee interested in driving sharp bargains, he becomes a kind of messiah and political evangelist with whom the author identifies himself more and more fully. The pattern of melodrama takes over and Morgan is made the bearer of all the hopes of humankind for escape from the tyranny, cruelty, and ignorance that have dominated the history of the race down to modern times. In this conception, the question of guilt cannot arise in connection with the Yankee because all evil is by definition embodied in the blood-stained and tear-soaked past. When he obliterates knights or priests with a dynamite bomb, he incurs no more guilt than do the caricature figures in an animated cartoon—not even at the end, when the gruesome Battle of the Sand Belt has left the Yankee and his band of loyal youths victorious but surrounded by twenty-five thousand decaying corpses of their enemies.[31]

There is thus no tragic flaw in the character of the Yankee Prometheus to account for his defeat. That defeat is inevitable in any case because Mark Twain has chosen to lay his fable in an actual country which obviously did not become an industrialized republic in the sixth century A.D. But even though the collapse of Morgan's program was implicit in the story from the beginning, the question of how Mark Twain rationalizes it is still important for the light it throws on his conception of progress. First the aristocracy and then the common people of Arthurian Britain are shown to be incapable of moving beyond the intellectual horizon established by their historical situation. This is the fundamental insight of the sociology of knowledge, but that science was still only embryonic in the 1880s. Mark Twain's term is "training." The discovery of the social rootedness of all thought became a landmark in his intellectual biography and threatened to dominate his last years. It seemed to him to mean that the human race is hopelessly damned.

The original premise of the contrast of civilizations in *A Connecticut Yankee* held that both the aristocracy of Arthur's Britain and the common people were locked into their feudal system by what a later day would call a false consciousness—that is, a systematic distortion of perception that rendered them unable to establish contact with social reality. But Hank Morgan, representative of the emancipated common people of nineteenth-century America, is assumed to be free of this intellectual blight. The situation is dramatized in the wonderfully amusing adventure of the enchanted pigsty, in which the Yankee accompanies the Demoiselle Alisaunde la Corteloise, or Sandy, to the rescue of what she believes to be forty-five noble ladies held in an enchanted castle by an ogre (Chap. 20). It will be recalled that Hank discovers the noble ladies to be a herd of swine, which he is able to buy from the swineherds for sixteen pence. A more serious version of false consciousness among the aristocracy had appeared somewhat earlier when the Yankee, now Sir Boss, paid a visit to the domains of Queen Morgan le Fay. The queen stabbed a page boy who by accident "fell lightly against her knee" (p. 196), and when the Yankee spoke to her of her "crime," she replied, "Crime! . . . How thou talkest! Crime, forsooth! Man, I am going to *pay* for him!" (p. 217).

This remark becomes the occasion for a curious digression in which the narrator expatiates on "training." "We have no thoughts of our own, no opinions of our own," he observes; "they are trans-

mitted to us, trained into us." And he develops the idea, with a
melancholy tone that is conspicuously in contrast with his usual
brisk self-assurance:

> All that is original in us, and therefore fairly creditable or discreditable
> to us, can be covered up and hidden by the point of a cambric needle,
> all the rest being atoms contributed by, and inherited from, a pro-
> cession of ancestors that stretches back a billion years to the Adam-
> clam or grasshopper or monkey from whom our race has been so
> tediously and ostentatiously and unprofitably developed. [p. 217]

This is not the occasion for attempting to unravel the varied strands
of thought that are entwined in this often-quoted sentence: it is
enough to note that the Yankee eventually adopts several different
forms of determinism (some emphasizing heredity, some environment), but the implication is plain that no educational program can
hope to overcome the inertia of perverted conditioning.

At the outset, however, the Yankee reformer is confident he can
overcome the bad training of the common people by bringing them
into functional contact with machines. He believes that exposure to
advanced technology will rid the workers of the "reverence" that
finds expression in abject loyalty to the two sources of oppressive
authority in the kingdom—the nobles and the Established Church.
In the end, however, he discovers that he has not "educated the
superstition out of those people" (p. 538). When the nobles, incited
by the Church, attack the Yankee, the mass of the population
supports "the 'righteous cause'"; the common people are not potential freemen and enlightened citizens of a republic, but "sheep" and
"human muck" (p. 551). This discovery is an explicit denial of the
hope implicit in the doctrine of progress; indeed, it seems to imply
that the contrast between the benighted South and the enlightened
North presented in the second part of *Life on the Mississippi* is
unchangeable. In other words, the outcome of *A Connecticut Yankee*
reveals a loss of faith in the doctrine of progress that was central to
the American sense of identity. The experience was so shocking that
Mark Twain's critics and even the writer himself were at first unable
to admit to consciousness the pessimistic implications of the ending
of the fable.

Contemporary readers, in so far as their opinions are recov-

erable, seem to have read the book as if Hank Morgan's revolution
had been successful. They virtually ignored the mountain of corpses
left by the Battle of the Sand Belt.[32] Even more astonishingly, they
do not seem to have noticed that at the end of the book the defeated
and dying protagonist, consumed by longing for his wife and child
left behind him in what he calls the Lost Land of sixth-century
Britain, is alone in an otherwise empty universe. There have been
hints from the beginning that the Arthurian world in which Hank
Morgan performs his many exploits, comic and heroic, and loses
both his final battle against the massed chivalry of England and his
personal contest with the rival magician Merlin, is a dreamland.
Like Tom Sawyer, enclosed in his hermetic world of fantasy, the
Yankee (with whom, incidentally, the author has by now identified
himself almost completely) seems to have dreamed the whole adventure.

The specter of solipsism haunted Clemens throughout the remaining twenty years of his life. It appears in many later works of fiction,
completed (like "My Platonic Sweetheart") or incomplete and
unpublished (like "Which Was the Dream?" and "The Great Dark"),
but is most fully developed in the long unfinished narratives from
which Albert B. Paine cobbled together the book he brought out
after Clemens's death as *The Mysterious Stranger*. Because the
writer's intellectual and emotional world was shattered during these
years, such fragmentary works were perhaps the most authentic
form of expression available to him. William M. Gibson, editor of
the *Mysterious Stranger Manuscripts*, points out that over the period
of seven years during which Mark Twain worked on them, "he also
composed a stream of notes and shorter pieces . . . concerning
some diabolic or angelic stranger."[33] Although the earliest version of
the Mysterious Stranger material begins as yet another sequel to
Adventures of Huckleberry Finn (it is laid in Hannibal, and Huck
and Tom are characters), the setting is soon moved to Austria at
some earlier period. The protagonist is always this new character,
called Little Satan or (inexplicably) "No. 44," or Philip Traum.[34]
Traits drawn from sources as diverse as Milton's Satan, Goethe's
Mephistopheles, and the boy Jesus in the Apocryphal New Testa-

ment are incorporated in the character,[35] but his most important lineage is to be found in the earlier transcendent figures in Mark Twain's own works, particularly Tom Sawyer, Colonel Sherburn, and Hank Morgan. In the final form of Satan this character achieves absolute transcendence. As an angel he is completely free of guilt—because like the Tom Sawyer who designed and produced the Evasion, he is beyond good and evil; in other words, as Mark Twain insists again and again, he lacks the Moral Sense that plagues mere human beings. Satan also has apparently unlimited intellectual and magical powers; he is omniscient, able to bring to life the clumsy clay figures molded by Huck and Tom (now called Niklaus and Theodor) and their friends, or to read the minds of other characters in the story, or to alter their destinies. But in exercising these supernatural powers he reveals his descent from Tom Sawyer (and ultimately, Mark Twain himself), for the world he controls proves in the end to be a dream world, and his control of it is merely the control of an artist over the world he imagines.

Even though the world in which Satan exists is unreal, however, the implication is that it is no more unreal than the historical world inhabited by the actual human race. Hank Morgan had declared this race to be "human muck" (p. 551). The collapse of the Yankee's program for redeeming man by means of progress and technological advance had left the writer abysmally disillusioned with mankind. In 1895 Mark Twain entered in his notebook his desire to write a book that would "scoff at the pitiful world, and the useless universe and the vile and contemptible human race. . . ."[36] This judgment, of course, assumes a perspective beyond the limitations of the merely human mind: it could be expressed in fiction only through the creation of a superhuman character. Thus Mark Twain identifies himself with the Mysterious Stranger, the angelic observer and playwright and stage manager. Only in this final phase is the inner logic of the impulse toward transcendence fully revealed. Satan explicitly denies the doctrine of progress that had been refuted by implication in the catastrophe at the end of *A Connecticut Yankee*. In "The Chronicles of Young Satan" (one of *The Mysterious Stranger* fragments) Satan mocks the idea on which Hank Morgan had based his program of reform for Arthur's Britain—the naïve belief that labor in mechanized factories would transform groveling serfs and slaves into enlightened citizens of a republic. Theodor Fischer, the boy narrator, says:

> In a moment we were in a French village. [Since Satan reveals the future as well as the past, this vision can represent literally any date; it must belong to the nineteenth century.] We walked through a great factory of some sort, where men and women and little children were toiling in heat and dirt and a fog of dust; and they were clothed in rags, and drooped at their work, for they were worn, and half-starved, and weak and drowsy. Satan said—
>
> "It is some more Moral Sense. The proprietors are rich, and very holy; but the wage they pay to these poor brothers and sisters of theirs is only enough to keep them from dropping dead with hunger. The work-hours are fifteen per day, winter and summer—from 5 in the morning till 8 at night—little children and all. And they walk to and from the pig-sties which they inhabit—four miles each way, through mud and slush, rain, snow, sleet and storm, daily, year in and year out. They get four hours of sleep. They kennel together, three families in a room, in unimaginable filth and stench; and disease comes, and they die off like flies."[37]

This passage, like many others in the book, makes social injustice the result of an ineradicable human trait. There is no prospect that mankind can be redeemed by history. The successive disillusionments for which Clemens's frightful personal calamities were, so to speak, the objective correlative led over time to the solipsism that had been implicit in Tom Sawyer's fantasies. It is spelled out in the final address of Satan to Philip Traum—a declaration appropriately left at Clemens's death as a floating fragment of manuscript, but faithful to the theme he had been struggling for decades to give its definitive expression:

> "It is true, that which I have revealed to you: there is no God, no universe, no human race, no earthly life, no heaven, no hell. It is all a Dream, a grotesque and foolish dream. Nothing exists but You. And You are but a *Thought*—a vagrant Thought, a useless Thought, a homeless Thought, wandering forlorn among the empty eternities!"[38]

Is this not a prophecy of the fictive world of Samuel Beckett or Thomas Pynchon? Stranger still, is not Satan's doctrine, in its literal meaning, also very close to that set forth by Emerson in *Nature* and "Self-Reliance"?

Henry James I:

SOWS' EARS AND SILK PURSES

Henry James began his literary career in the 1860s by reviewing books, especially fiction, for the *North American Review*, the *Atlantic Monthly*, and the *Nation*. William Veeder has recently listed more than seventy current English and American novels James is known to have read before 1880, with the assurance that he read many more whose titles are not recorded. Most of this fiction, Veeder points out, especially that having the widest sale, offers the excitement of "hairbreadth escapes, touching deaths, [and] violent confrontations," yet at the same time reassures the reader that "God is in His heaven or at least that order remains on earth."[1] James's attitude toward it alternates between patronizing amusement and caustic dismissal.[2] Yet in diction, characters, and plots James's own early work relies heavily on the same resources as do the current bestsellers. The story of the writer's development is the story of how this "apprentice, steeped in popular practice, moves hesitatingly and with lapses to those transformations which will alter the form of fiction."[3] He manages eventually to free himself from conventional attitudes and the diction in which they are expressed—mainly by achieving an ironic perspective on them.

On the assumption that Veeder's thesis needs no further support, I propose to consider other aspects of James's relation to popular culture, especially as it was embodied in the novels that can be called middlebrow or lowbrow. By 1880, in *Washington Square*, James was able to make effective use of the "taste for light literature" that he attributes to Dr. Sloper's widowed sister, Mrs. Penniman, to whom after the death of his wife the Doctor entrusts the education of his only child Catherine.[4] It soon becomes apparent that the fantasies Mrs. Penniman has garnered from her reading have made her into not only a foolish but also a dangerous woman by rendering her incapable of perceiving the realities of human character. The narrative voice announces that Mrs. Penniman "would have liked to have a lover, and to correspond with him under an assumed name, in letters left at a shop" (p. 13). She inflicts an irreparable injury on Catherine by encouraging the inexperienced girl to form an exaggerated estimate of the devotion and sincerity of Morris Townsend, who is subsequently revealed to be interested only in the fortune he believes Catherine will inherit from her father.

James describes Mrs. Penniman's attitude in elaborate metaphors drawn from the popular theater:

> [She] delighted of all things in a drama, and she flattered herself that a drama would now be enacted. Combining as she did the zeal of the prompter with the impatience of the spectator, she had long since done her utmost to pull up the curtain. She, too, expected to figure in the performance—to be the confidante, the Chorus, to speak the epilogue. It may even be said that there were times when she lost sight altogether of the modest heroine of the play in the contemplation of certain great scenes which would naturally occur between the hero and herself. [p. 81]

James returns to the theme later in an even more elaborate metaphor, drawn this time from the fictional equivalent of melodrama:

> Mrs. Penniman's real hope was that the girl would make a secret marriage, at which she should officiate as bride'swoman or duenna. She had a vision of this ceremony being performed in some subterranean chapel; subterranean chapels in New York were not frequent, but Mrs. Penniman's imagination was not chilled by trifles; and of the guilty couple—she liked to think of poor Catherine and her suitor as the guilty couple—being shuffled away in a fast-whirling vehicle to some obscure lodging in the suburbs, where she would pay them (in a thick veil) clandestine visits; where they would endure a period of

romantic privation; and [where] ultimately, after she should have been their earthly providence, their intercessor, their advocate, and their medium of communication with the world, they would be reconciled to her brother in an artistic tableau, in which she herself should be somehow the central figure. [p. 126]

There is more imagery of this kind in the novel—James obviously enjoyed it—but the quoted passages are enough to reveal the important features of the clichés he means to satirize. The fantasies Mrs. Penniman has taken over ready-made from "light literature" are dangerous because they support the basic delusion of popular culture: the notion that one can eat his cake and have it too.[5] Mrs. Penniman dreams of a future in which Catherine and Morris will have the romantic thrill of defying Dr. Sloper, yet also enjoy the solid financial benefits of remaining in his good graces. The matters that James is most interested in—the quality of Dr. Sloper's feeling for Catherine, and of hers for him; Morris's pretended and her genuine love; the two men's mistaken perceptions of the girl—these matters hardly figure in Mrs. Penniman's estimate of the situation because she has borrowed a complete set of the appropriate emotions from her literary or theatrical models. Once Catherine and Morris are married, the literary conventions call for Dr. Sloper to relent, and the happiness of the young lovers will be assured by the nobility of Morris's feelings despite Catherine's cold unemotional nature.

When Morris becomes convinced that Dr. Sloper really will disinherit Catherine if she marries him, and decides therefore to withdraw, he invents an excuse to cover his retreat that is entirely to Mrs. Penniman's taste: "You can explain to her why it is. It's because I can't bring myself to step in between her and her father— to give him the pretext he grasps at so eagerly (it's a hideous sight!) for depriving her of her rights." Mrs. Penniman is charmed by "this formula." "That's so like you," she exclaims; "it's so finely felt" (p. 228). Morris's reply is eloquent: "Oh damnation!" (He has concluded long before (p. 131) that "The woman's an idiot!")

James did not use again the typical distortions of popular literature for so important a function in a novel but through the years he turned to material of this kind for a variety of other pur-

poses. Two examples must suffice. In *The Princess Casamassima*
(1886) James makes the magnificent Cockney girl Millicent Hen-
ning a foil for Christina Light, the beautiful half-American *princi-
pessa* who has joined the anarchist conspiracy that also recruits
young Hyacinth Robinson, apprentice to a bookbinder and ille-
gitimate son of an English nobleman and a French woman of the
working class. Hyacinth has aristocratic tastes in literature, in painting, in women's dress, etc., that James implies are inherited from his
father,[6] and the point comes up frequently because Hyacinth's esthetic judgments are at variance with Milly's. James establishes the
contrast between Hyacinth's and Milly's tastes by many small
touches, but a sampling will have to serve. Thus Hyacinth, as a child,
reads the opening pages of "romances" in story papers (the *Family
Herald* and the *London Journal*) that are displayed on news-stands,
but as an adult he turns to Browning's *Men and Women*. Milly,
however, continues to enjoy the story-papers.[7] At her insistence,
Hyacinth manages to get passes to a melodrama entitled *The Pearl
of Paraguay* that is having a long run at the Strand theater (1:180).
The action of the play unfolds itself "through scenes luxuriantly
tropical, in which the male characters wore sombreros and stilettos
and the ladies either danced the cachucha or fled from licentious
pursuit. . . . "[8] Milly is much more amused than he is by the
"horse-play" of the farce that serves as a "traditional prelude" to the
play (1:188). In the third act, Milly is

> moved to tears . . . when the Pearl of Paraguay, dishevelled and
> distracted, dragging herself on her knees, implored the stern hidalgo
> her father to believe in her innocence in spite of circumstances ap-
> pearing to condemn her—a midnight meeting with the wicked hero
> in the grove of cocoanuts. [1:188–190]

Hyacinth recognizes this is flimsy stuff, and is instantly aware that
the sophisticated Captain Sholto, a man-about-town, certainly no
esthete but of higher social rank, does not "take the play seriously"
(1:198).

Yet Hyacinth loves the theater; one can almost imagine that
James is describing his own reactions when he says of the youth
(without relation to the requirements of the plot):

. . . the theatre, in any conditions, was full of sweet deception for him. His imagination projected itself lovingly across the footlights, gilded and coloured the shabby canvas and battered accessories, losing itself so effectually in the fictive world that the end of the piece, however long or however short, brought with it something of the alarm of a stoppage of his personal life. It was impossible to be more friendly to the dramatic illusion. [1:188]

Later, James uses responses to the music-hall stage to demonstrate a similar difference of cultural levels by noting that Paul Muniment, the able anarchist leader,

excited on Hyacinth's part a kind of elder-brotherly indulgence by the open-mouthed glee and credulity with which, when the pair were present, in the sixpenny gallery, at Astley's, at an equestrian pantomime, he followed the tawdry spectacle. [1:228]

From this perspective, the tawdriness of popular art could seem merely an element of local color, like the Cockney dialect of Milly and other natives of the London slums which James renders with unexpected care (although without notable felicity). But he also continued to use popular fiction as he had used it in portraying Mrs. Penniman in *Washington Square*, to show how it distorts perceptions. The heroine of *In the Cage* (1898), whose membership in the anonymous lower middle-class is emphasized by the fact that she is never given a name, reads novels that she gets from a rental library, "very greasy, in fine print and all about fine folks, at a ha'penny a day."[9] Her job (in that era before the telephone came into general use) is to receive telegrams for transmission in a branch post-office that occupies crowded space in a grocery store near an elegant apartment house in the West End of London. The exalted notions of how the members of the upper classes deport themselves that she has gained from her novels lead her to develop elaborate fantasies about the meaning of the messages that pass through her hands, especially those sent by one Captain Everard. James suggests the texture of the situation she imagines for him by a kind of represented discourse employing the strained, hyperbolic imagery (although not the diction) of the girl's favorite reading matter:

> He was . . . in the strong grip of a dizzy splendid fate; the wild
> wind of his life blew him straight before it. Didn't she catch in his face
> at times, even through his smile and his happy habit, the gleam of that
> pale glare with which a bewildered victim appeals, as he passes, to
> some pair of pitying eyes? [p. 69]

The girl naturally conceives a role for herself in this drama:

> She quite thrilled herself with thinking what, with such a lot of mate-
> rial, a bad girl would do. It would be a scene better than many in her
> ha'penny novels, this going to him in the dusk of evening [when she
> gets off work] at Park Chambers and letting him at last have it. "I
> know too much about a certain person now not to put it to you—excuse
> my being so lurid—that it's quite worth your while to buy me off.
> Come, therefore; buy me!" There was a point indeed at which such
> flights had to drop again—the point of an unreadiness to name, when
> it came to that, the purchasing medium. It wouldn't certainly be
> anything so gross as money, and the matter accordingly remained
> rather vague, all the more that *she* was not a bad girl. [pp. 67–68]

In the end, of course, the illusion is dispelled. The Captain is
revealed as a rather tawdry adventurer, and the girl in the cage
resigns herself to the marriage with a rising grocer's clerk that she
has been tacitly resisting for a long time.

James's urbane amusement at the circus-poster effects of definitely
lowbrow fiction does not suggest intimate acquaintance with it:
in the prefaces to the New York Edition he refers only vaguely once
or twice to stories involving "the surprise of a caravan or the
identification of a pirate," or the adventures of "detectives or pirates
or other splendid desperadoes."[10] But his dependence on the monthly
magazines to serialize his novels made it impossible for him to
maintain such a lofty distance from the canons of middlebrow taste.
The requirement of a happy ending is a case in point. In 1876,
when Howells had accepted James's *The American* for serial publication in the *Atlantic,* he tried to persuade his friend to have
Claire de Cintré marry Christopher Newman at the end.[11] James
rejected the suggestion, and later he explained his reasons at length:

I quite understand that as an editor you should go in for "cheerful endings"; but I am sorry that as a private reader you are not struck with the inevitability of the American dénouement. I fancied that most folks would feel that Mme de Cintré *couldn't*, when the finish came, marry Mr. Newman; and what the few persons who have spoken to me of the tale have expressed to me (e.g. Mrs. Kemble t'other day) was the fear that I should really put the marriage through. *Voyons;* it would have been impossible: they would have been an impossible couple, with an impossible problem before them. For instance—to speak very materially—where would they have lived? . . . No, the interest of the subject was, for me, (without my being at all a pessimist) its exemplification of one of those insuperable difficulties which present themselves in people's lives and from which the only issue is by forfeiture—by losing something.[12]

But James promised to try to meet Howells's desires in the future, and in *The Europeans,* serialized in the *Atlantic* two years later, he did so. He explained to Elizabeth Boott: "The offhand marrying in the end was *commandé* . . . it had been a part of the bargain with Howells that *this* termination should be cheerful and that there should be distinct matrimony. So I did [hit] it off mechanically in the closing paragraphs."[13]

It should be emphasized that Howells was not idiosyncratic in his editorial concern for the sensibilities of subscribers to the magazines. As late as 1896 Robert U. Johnson, editor of the *Century,* which had serialized *The Bostonians* and other novels by James, rejected an essay on Dumas *fils* that had been explicitly commissioned, on the grounds that James mentioned Dumas' fondness for the theme of seduction.[14] James caricatured the incident (shifting the scene to London) in "John Delavoy" (1898), in which the editor of *The Cynosure* rejects a critical article on an obscure but accomplished novelist with the dogmatic announcement:

"You're not writing in *The Cynosure* about the relations of the sexes. With those relations, with the question of sex in any degree, I should suppose you would already have seen that we have nothing whatever to do. If you want to know what our public won't stand, there you have it."[15]

So far as is known all James's writings eventually found a market (except *The Ivory Tower,* left unfinished at his death). But he was

never able to feel entirely secure, and with the passage of time
his difficulties with editors and publishers increased rather than
diminished. From the middle 1880s onward he expressed himself on
this topic at intervals with remarkable bitterness. A letter to Howells
from Paris in 1884 contains a passage that could be duplicated from
other letters of that and later periods:

> What you tell me of the success of ————'s last novel [Percy
> Lubbock, the editor of the *Letters*, discreetly suppresses the name of
> the writer in question] sickens and almost paralyzes me. It seems to
> me (the book) so contemptibly bad and ignoble that the idea of people
> reading it in such numbers makes one return upon one's self and ask
> what is the use of trying to write anything decent or serious for a public
> so absolutely idiotic. It must be totally wasted. . . . Work so shame-
> lessly bad seems to me to dishonour the novelists's art to a degree that
> is absolutely not to be forgiven; just as its success dishonours the
> people for whom one supposes one's self to write. Excuse my ferocities,
> which (more discreetly and philosophically) I think you must share;
> and don't mention it, please, to anyone, as it will be set down to green-
> eyed jealousy.[16]

Comments of similar purport bear out James's disappointment
at the reception of *The Bostonians* and *The Princess Casamassima*,
both published in 1886. In January 1888 he wrote to Howells:

> I have entered upon evil days—but this is for your most private ear. It
> sounds portentous, but it only means that I am still staggering a good
> deal under the mysterious and (to me) inexplicable injury wrought—
> apparently—upon my situation by my two last novels, the *Bostonians*
> and the *Princess*, from which I expected so much and derived so little.
> They have reduced the desire, and the demand, for my productions to
> zero—as I judge from the fact that though I have for a good while past
> been writing a number of good short things, I remain irremediably
> unpublished. Editors keep them back, for months and years, as if they
> were ashamed of them, and I am condemned apparently to eternal
> silence.[17]

The declining sales of James's novels led him to abandon that
form temporarily in favor of shorter fictions. Far more significant,
however, was his much-discussed effort to write for the theater. This
undertaking proved to be singularly ill-advised: it is hard to imagine
a talent less adapted to dramatic writing than James's. Furthermore,

the state of the drama in London as well as in New York in those
years was notoriously degraded. James wrote apologetically to Robert
Louis Stevenson in 1891, *"Je fais aussi du théâtre,"* and added:

> Don't be hard on me—simplifying and chastening necessity has laid
> its brutal hand on me and I have had to try to make somehow or other
> the money I don't make by literature. My books don't sell, and it looks
> as if my plays might. Therefore I am going with a brazen front to write
> half a dozen.[18]

Two years later James wrote to Stevenson again, with less confidence:

> . . . I am working with patient subterraneity at a trade which it is
> dishonour enough to practise, without talking about it. . . . the
> *book*, as my limitations compel me to produce it, doesn't bring me in a
> penny. . . . I *don't* sell ten copies!—and neither editors nor pub-
> lishers will have anything whatever to say to me.[19]

After several years of effort, culminating in months of frustrating
attendance at rehearsals, James suffered what he considered a public
disgrace when, on the opening night of his play *Guy Domville* in
1895, the polite applause from the boxes and stalls was drowned out
by jeers and boos from the gallery. It was a humiliation he could
never forget.[20]

A shock of this magnitude could hardly fail to register itself on
James's imagination; one would expect it to be recorded in some
form in his work. Such indeed was the case. For more than a decade
James's shorter fictions display from time to time one or another
component of a constellation of images evidently generated by the
"evil days" during the later 1880s and 1890s. There is always a writer
of genius whose work is neglected by the world but appreciated by a
small minority of sensitive readers. Often there is also the contrasting figure of an immensely popular novelist, usually female,
whose books have absolutely no literary merit. In handling these
materials James abandons his usual restraint and balance, resorting
to overstrained diction that approaches the melodramatic style of the
vulgar fiction and drama he had often satirized. From a half-dozen

tales that develop one aspect or another of this contrast, I shall take a closer look at two.

"Greville Fane" (1892) has for its central figure Mrs. Stormer, a woman novelist who through a long and lavishly productive career has used a male pseudonym. She is not an entirely unsympathetic character—merely "a dull, kind woman." The male narrator, a dozen years younger, has a condescending fondness for her: "This is why I liked her," he explains, "—she rested me so from literature."[21] In depicting Mrs. Stormer, James seems fascinated by the possibility of a gift for story-telling that has nothing to do with either art or ideas. The narrator offers a thumbnail account of Greville Fane's character and her career:

> To myself literature was an irritation, a torment; but Greville Fane slumbered in the intellectual part of it like a Creole in a hammock. She was not a woman of genius, but her faculty was so special, so much a gift out of hand, that I have often wondered why she fell below that distinction. This was doubtless because the transaction, in her case, had remained incomplete; genius always pays for the gift, feels the debt, and she was placidly unconscious of obligation. She could invent stories by the yard, but she couldn't write a page of English. She went down to her grave without suspecting that though she had contributed volumes to the diversion of her contemporaries she had not contributed a sentence to the language. This had not prevented bushels of criticism from being heaped upon her head; she was worth a couple of columns any day to the weekly papers, in which it was shown that her pictures of life were dreadful but her style really charming. [8:436]

James's fascinated portrayal of Greville Fane includes a description of her work that reveals his conception of the kind of novel to be found on the lists of best-sellers. The writer he has most in mind is perhaps Ouida, although Mary Elizabeth Braddon is also a possible model:

> An industrious widow, devoted to her daily stint, to meeting the butcher and baker and making a home for her son and daughter, from the moment she took her pen in hand she became a creature of passion. She thought the English novel deplorably wanting in that element, and the task she had cut out for herself was to supply the deficiency. Passion in high life was the general formula of this work, for her imagination was at home only in the most exalted circles. She adored,

in truth, the aristocracy, and they constituted for her the romance of the world or, what is more to the point, the prime material of fiction. [8:437]

When the narrator, a novelist himself whose work, wrought in anguish, has only negligible sales, expounds to Greville Fane the indispensability of form, she declares it "a pretension and a *pose*." For his part, he compares her to "a common pastry-cook, dealing in such tarts and puddings as would bring customers to the shop." Pressing the charge, he says that "She had a serene superiority to observation and opportunity which constituted an inexpugnable strength and would enable her to go on indefinitely." This description leads the narrator (apparently speaking for James) to frame an epigram that is both cynical and false: "It is only real success that wanes, it is only solid things that melt" (8:438). Finally, when the narrator declares, "I try, in my clumsy way, to be in some direct relation to life!" Greville Fane replies, "Oh, bother your direct relation to life!" (8:439).

The plot of this tale is of course too schematic; it falls far below the level of James's best work, but in its clumsy way it enforces his moral case against the kind of conventional fiction Greville Fane produces. She has devoted herself to providing for her offspring what she considers the highest attainable goods in life: a conventionally brilliant marriage for her daughter, and for the son careful training in the craft of writing successful fiction. The daughter has become a petty snob discontented with her situation as the wife of a marginal politician; the son has learned no art save that of sponging on his mother for funds to support his empty career as man-about-town. One is tempted to conclude that James's mind, however fine, could be violated by an idea when the idea was that of the immorality of best-selling fiction.[22]

"The Next Time" (1895) grew out of a notebook entry that James made within a month after the *Guy Domville* fiasco:

The idea of the poor man, the artist, the man of letters, who all his life is trying—if only to get a living—to do something *vulgar*, to take the measure of the huge, flat foot of the public: isn't there a little story in it, possibly, if one can animate it with an action; a little story that might perhaps be a mate to *The Death of the Lion?* It is suggested to

me really by all the little backward memories of one's own frustrated
ambition—in particular by its having just come back to me how,
already 20 years ago, when I was in Paris writing letters to the *N.Y.
Tribune*, Whitelaw Reid wrote me to ask me virtually *that*—to make
'em baser and paltrier, to make them as vulgar as he could, to make
them, as he called it, more "personal." Twenty years ago, and so it
has been ever, till the other night, Jan. 5th, the *première* of *Guy
Domville*. Trace the history of a charming little talent, charming
artistic nature, that has been exactly the martyr and victim of that
ineffectual effort, that long, vain study to take the measure above-
mentioned, to "meet" the vulgar need, to violate his intrinsic conditions . . .[23]

James goes on to wonder whether he might not oppose to this artist
"some contrasted figure of another type—the creature who, dimly
conscious of deep-seated vulgarity, is always trying to be refined,
which doesn't in the least prevent him—or her—from succeeding.
Say it's a woman. . . . " And it turned out to be a woman, Mrs.
Highmore, author of eighty volumes of best-selling fiction, whom
James associates in a notebook entry with the perennially successful
Mary Elizabeth Braddon.[24]

The incident involving James's letters to the New York *Tribune*
in 1876 has been explored by Ilse Dusoir Lind. Reid had asked him
to strive for more "brevity, variety, and topics of wide interest" in his
dispatches, in place of "literary treatment" of subjects.[25] James
retorted to Reid that the articles he had been sending were "the
poorest I can do, especially for the money." His anger seems dispro-
portionate, although perhaps understandable in view of his recent
disappointment. But to the end of his life James continued to drop a
remark occasionally implying that an immensely popular novel is
simply a clumsy effort to write a good one. In "The Next Time" he
says of Mrs. Highmore:

It was not that when she tried to be what she called subtle . . . her
fond customers, bless them, didn't suspect the trick nor show what they
thought of it: they straightway rose on the contrary to the morsel she
had hoped to hold too high, and, making but a big, cheerful bite of it,
wagged their great collective tail artlessly for more. It was not given
to her not to please, nor granted even to her best refinements to
affright.[26]

However childish James's resentment may seem in this passage, the image of the public as dog is undeniably amusing.

Out of James's bitterness he makes an apologue: opposite the older woman who is unable to produce anything that does not sell he introduces her son-in-law, Ray Limbert, a young novelist of genius who cannot produce anything that does. Having established this mechanical symmetry, James begins playing games with it. Limbert receives a commission like James's commission from the *Tribune* to write a "London Letter" for an Edinburgh paper, and he valiantly undertakes to produce the kind of "trash" that is required. His effort is described as grinning through a horse-collar (9:195, 197). The plot takes a comic turn and grows extremely complex. Limbert's friends hope anxiously that he may succeed in writing vulgarly enough to satisfy his employers; and he succeeds, temporarily, to such an extent that he is placed in charge of a literary magazine which the owner, Mr. Bousefield, thinks he wishes to have maintain a high literary standard. Limbert plans a deception faintly reminiscent of Melville's notion of aiming *Pierre* simultaneously at two distinct audiences—one large and stupid, another small and perceptive.[27] There is, however, no reason to undertake a plot summary. What one remembers from this story is several moments of great eloquence that evidently derive their force from James's conception of his own situation. Thus poor Limbert declares: "I want to sell. . . . I must cultivate the market—it's a science like another. . . . I haven't been obvious—I must *be* obvious. I haven't been popular—I must *be* popular. It's another art—or perhaps it isn't an art at all. It's something else; one must find out *what* it is" (9:208). Yet the mystery of Limbert's mother-in-law's unfailing appeal to her vast audience continues to baffle him. "She can't explain herself much," he declares: "she's all intuition; her processes are obscure; it's the spirit that swoops down and catches her up. But I must study her reverently in her works. . . . my loins are girded: I declare I'll read one of them—I really will: I'll put it through if I perish" (9:208). All in vain: the narrator indulges in an almost macabre cadenza on the topic of the failure of Limbert's books to sell:

> Several persons admired his books—nothing was less contestable; but they appeared to have a mortal objection to acquiring them by

subscription or by purchase: they begged or borrowed or stole, they delegated one of the party perhaps to commit the volumes to memory and repeat them, like the bards of old, to listening multitudes. [9:207]

In a similar vein is a later passage in which the narrator tells Mrs. Highmore that another of Limbert's books "has an extraordinary beauty," and she exclaims: "Poor duck—after trying so hard!" (9:219).

From such comic fantasy, however, James oscillates in this odd tale to moments of intense seriousness. The narrator, transparently the author's mouthpiece, describes a novel Limbert is writing as ". . . that fiery-hearted rose as to which we watched in private the formation of petal after petal and flame after flame" (9:198). The narrator sits up all night reading yet another novel, intended by Limbert to be "the worst he could do," only to discover that "the perversity of the effort, even though heroic, had been frustrated by the purity of the gift."

> Indeed [continues the narrator] as the short night waned and, threshing about in my emotion, I fidgeted to my high-perched window for a glimpse of the summer dawn, I became at last aware that I was staring at it out of eyes that had compassionately and admiringly filled. The eastern sky, over the London housetops, had a wonderful tragic crimson. That was the colour of his magnificent mistake. [9:213]

Many passages have a strong autobiographical suggestiveness. The narrator declares, for example, that although Limbert will try "for the common," the result will turn out to be "only his genius in an ineffectual disguise" (9:220). Limbert's declared effort to write a mediocre and therefore popular novel—"the effort that had been the intensest of his life" (9:223)—results in a book that annoys the publisher of the magazine Limbert has been hired to edit because it makes excessive intellectual demands on the readers and does not "allow for human weakness" (9:216). The novelist is dismissed from his editorial post. This experience is said to have been a turning point in Limbert's career: "a spring had really snapped in him" (9:223). And Limbert's loyal wife has attitudes resembling those expressed from time to time by James himself: ". . . she would almost have been ashamed of him if he had suddenly gone into editions . . . She would have liked the money immensely, but she

would have missed something she had taught herself to regard as
rather rare" (9:227). The basic idea underlying "The Next Time"
has every sign of being James's conception of his own predicament.
Limbert can not be vulgar even when he tries: "It takes more than
trying" says the narrator: "—it comes by grace. It happens not to be
given to Limbert to fall." In short: ". . . you can't make a sow's
ear of a silk purse" (9:220).

8

Henry James II:
THE PROBLEM OF AN AUDIENCE

The financial hardships suffered by the neglected geniuses in James's later stories, while they do not approach actual poverty, are nevertheless more drastic than James himself had to endure. So far as I know, an exact computation of his income has never been published. Some of the people he mingled with socially were extremely wealthy, and he may have been led to exaggerate his own financial difficulties by contrast with their affluence. He was able to buy Lamb House in Rye on the Sussex coast, and to maintain several servants there (as well as to keep a room permanently in his London club for use on his visits to town). Edel estimates James's income 1908-1910 at between $5000 and $6500 a year (equivalent, say, to $25,000 in our own day?)—certainly a comfortable amount for a bachelor without dependents.[1] Yet there is no doubt that James often felt hard-pressed financially, and he did a prodigious amount of occasional writing for magazines for the sake of the income it brought him. Edith Wharton, who knew as much as anyone about his private affairs, took James's financial troubles seriously enough to propose a subscription for his benefit among her acquaintances, and when James flatly rejected this notion, she managed in 1912 to transfer $8000 of her own royalties secretly to his account with Scribners.[2]

In addition to imagining himself to be nearer bankruptcy than he was, James seems to have held an exaggerated notion of his obscurity. The subtle novelists in "John Delavoy" and "The Next Time," for example, not only find few readers but are totally unknown to the public. On the other hand, from the 1880s onwards James himself was consistently rated as a major writer despite the meager sales of his books. In balloting by readers of the *Critic* in 1884 to designate members of a proposed American Academy of Letters he ranked thirteenth. (Howells placed fifth and Mark Twain fourteenth.) In the same year James ranked tenth in a similar poll conducted by the Harvard *Herald-Crimson*—above both Howells, in fourteenth place, and Mark Twain in twentieth. A poll conducted by *Literature* in 1899 gave Howells first place and Mark Twain third place, but James had also advanced, to sixth place.[3]

These evidences of public esteem, however, and the financial support offered him by editors of magazines, were outweighed in his mind by the dwindling sales of his books. In 1915, only a few months before his death (and, to be sure, a year after the outbreak of the First World War), he told Edmund Gosse that the sale of his books was virtually at a standstill. The New York Edition, he said, had been in effect a catastrophe: ". . . my annual report of what it does—the whole 24 vols.—in this country amounts to about £25 from the Macmillans; and the ditto from the Scribners in the U.S. to very little more."[4] Given these circumstances, a more detailed examination of what reviewers had to say about James's work promises to be both interesting and enlightening. I shall therefore undertake a brief survey of the critical reception accorded his books from the 1880s to his death more than thirty years later, with particular attention to the stages through which his reputation passed. Although I have not been able to carry out a quantitative content analysis, I have read many reviews in both English and American journals. The summary that follows is based on an intuitive selection of passages that seem to me to embody representative attitudes toward his work.

During James's apprenticeship as a novelist, the attention of American writers, critics, and readers alike was fixed on the debate over "idealism" versus "realism."[5] This chapter of literary history has been abundantly documented; there is no point in going over it here again in any detail, except for the sake of attacking some apparently indestructible confusions and obscurities. Two kinds of novel commanded wide audiences during the middle decades of the nineteenth century: the sentimental story of courtship or family life, and the "sensation" novel (sometimes faintly scandalous), developing "dramatic" conflicts and "passion." Both were basically melodramatic, and both invited identification of the reader with heroes and heroines of exceptional beauty and actual or metaphorical nobility.[6] James, on the other hand, along with his friend Howells, was soon recognized as a leading practitioner of "the new novel," which Howells undertook to define in a controversial essay in the *Century* in 1882.[7] These two young writers were popularly understood to produce books departing so drastically from the familiar formulas that many reviewers insisted on calling them "studies" rather than novels. The phrase was sometimes expanded to "studies of character"[8] but the kind of study implied was quite unlike traditional character-drawing, say, in Dickens. The method of the fictional study was assumed to be that of "psychological" (that is, scientific) analysis.[9]

Reading the reviews of James's and Howells's novels during the 1870s, one gets the impression that a majority of critics and doubtless an even greater majority of readers resisted the new analytical fiction. It demanded much more exertion from the reader than did the conventional novel, because it deliberately refused to provide channels for the discharge of crude emotion that had become institutionalized in earlier decades. The new novel took itself seriously and demanded that its readers do the same.[10] It refused to offer itself as merely an amusement or as a sentimental debauch. Instead, just as Margaret Oliphant had castigated Hawthorne in the 1850s, reviewers complained again and again that James insisted on introducing them to the dissecting room of the anatomist, with repulsive associations of blood and stench and callousness. When "dissection" becomes "vivisection," the notion of callousness acquires overtones of sadism. Even when the "study" is linked merely with the notations that a graphic artist brings home in his sketchbook from a walk, it is accused by implication of lacking a solid structure or finished surface. Furthermore, psychological analysis, as contrasted with dramatic conflict and action, was likely to be boring. For example, Mrs. Sarah Butler Wister declared in the *North American Review* that

> the wonderfully told vicissitudes of feeling [of James's characters] leave us cold; it is not that they are unlike real people; they are most real and living, but we do not identify ourselves with them; we never for a moment cease to be spectators; we are intellectually interested, but as unmoved as one may suppose the medical class of a modern master of vivisection to be.

And later she returns to the charge: "The effect of this perpetual analysis is fatiguing; the book never ceases to interest, but it taxes the attention like metaphysics."[11]

When James called *The Europeans* a "sketch" on the title page, according to Richard Grant White, reviewer for the *North American Review*, the term meant that James himself recognized the work lacked a plot, and that "in writing it he did not propose to himself to interest his readers strongly in the fate of his personages."[12] An anonymous reviewer of *Washington Square* for the *Californian* repeated the standard opinions that James "analyzes too much" and treats his characters as if he were performing "a difficult vivisection," so that "we feel that the piercing of live flesh in cold blood is bad art. . . ."[13] The metaphor of dissection recurs frequently in discussion of *The Portrait of a Lady*. James H. Morse declares that the analytical or scientific method prevents enthusiastic attachment to any character and thus tends "to eliminate the spiritual action of the imagination."[14] The reviewer for the *Literary World* goes so far as to say that the *Portrait* "might almost be called a cruel book in its dissection of character and exposure of the nerves and sinews of human actions."[15]

The contrast between "dramatic" and "analytical" or "scientific" fiction was identical with the contrast between "idealism" and "realism" and would continue to be agitated in the pages of literary magazines for another twenty years. The debate does not seem very interesting now because neither side appears to have been able to grasp what the significant issues were. From the perspective of our own day we can recognize that the "idealists" were simply defending the kind of middlebrow sentimental fiction that had become popu-

lar in the 1850s, while the "realists" were in some respects a kind of
avant-garde attempting to domesticate in the United States the new
fictional mode invented in France by the generation after Balzac.[16]
The attitudes that hostile critics called "scientific" were derived
from the tendency toward secularization that was visible on both
sides of the Atlantic: Julian Hawthorne was correct in maintaining
that literary realism represented a spirit of agnosticism.[17] Even when
the traditional novel had abandoned the aggressive evangelical piety
of *The Wide, Wide World*, or *The Lamplighter*, or *Uncle Tom's
Cabin*, it remained highly moralistic; readers had learned to expect
to be able to identify themselves with characters portrayed as being
without moral blemish. Confrontation between virtuous and wicked
characters inevitably gave rise to "dramatic" (that is, sensational)
situations. And since the moral principles or "laws" that were
believed to control the universe were identical with those incorporated in the dominant value system of American society, "idealistic" stories provided powerful reassurance to the reader, lessening
anxiety by allowing him (more often her) to project guilt into a
hated villain (or, more rarely, the female equivalent). The happy
ending that was a routine requirement in popular fiction confirmed
the reader's self-righteousness further by completing the distribution of rewards and punishments according to generally accepted
moral criteria.

The Bostonians, published in 1886, aroused much more hostility in reviewers than had *The Portrait of a Lady*. Part of the
difficulty was a local scandal caused by the resemblance of Miss
Birdseye to Elizabeth Palmer Peabody, Hawthorne's sister-in-law;[18]
but this affected only a small group of New Englanders, whereas the
negative reviews were widely distributed. James said later that Gilder, editor of the *Century*, in which the novel was serialized, told
him "they had never published anything that appeared so little to
interest their readers."[19] To the by then habitual complaints of
reviewers about James's excessive analysis was added a new objection to the descriptive detail which he had added to *The Bostonians*
in a deliberate effort to try out the method of Zola and other French
naturalists. William Morton Payne combined the two charges in
declaring that the author had been "wearisomely minute in his
. . . analysis" and had cluttered the story with "a mass of analysis

of trifling things."[20] Even Annie R. M. Logan, usually well-disposed toward James, declared (somewhat ambiguously) in the *Nation* that the book contained "a mass of what, from a hasty reading, may be stigmatized as super-subtle analyses, ultra-refined phrases, fine-spun nothings."[21] It is therefore not surprising that the *Independent*, considerably lower on the brow scale than the *Nation*, should have seized the occasion to emphasize the stratification of the audience for fiction. The self-righteous tone is almost a defining trait of populist anti-intellectualism:

> In our museums and art-collections there are special departments and galleries devoted to, let us say, Aztec relics, cuneiform inscriptions, the evidences of the stone age, and the like. That these matters are set forth by themselves in clean and very quiet corners, does not mean that nobody in the world takes pleasure in inspecting them, or likes to set eyes on them. Not so. It merely signifies that most of us don't. It is implied that to the general, but by no means uneducated, public the spending an hour or so before those crowded shelves would be an insufferable bore, and make life a burden for the whole afternoon. Very good: Mr. James may nowadays be looked upon as the head of a certain college of savants, a man delighting in writing what to the majority of flesh-and-blood men and women has no excuse for being so praised, and books that grow more and more dull. This long, prosy, carefully written novel was not worth writing and is unreadable. Only a certain class of readers will be able to honestly say or think that they admire it or find anything in it that takes hold on them. . . . The rest of us will know enough to walk quickly by, not blaming the devotees to archeology and the stone age for their taste, deeply humiliated that we prefer other mental material, but also honest enough to say so.[22]

On the same page another unsigned review (by the same critic?) praises Mrs. Amelia E. Barr's *The Last of the MacAllisters* (apparently Scott-and-water—the reviewer actually mentions "the Man of Abbotsford") for its "rugged strength and picturesqueness."

Some hostility, or at least indifference, toward James was to be expected in the *Independent*. But the degree of hostility in this brief notice is not easy to account for. Since Olive Chancellor is the character whose mental processes are portrayed in greatest detail in the novel, the reviewer may be disturbed by the fact that James presents here the first conspicuous treatment of lesbianism as a

motivating force in American fiction. The suggestion of self-contra-
diction in objecting simultaneously to analysis (that is, careful
depiction of states of mind) and excessive detail (that is, careful
attention to outer circumstances) may mean that the reviewer has
not managed to identify and express his real objection. There may
also be some significance in the slightly more favorable treatment
accorded to *The Princess Casamassima* apparently by the same
reviewer only a few months later. The gravamen of his complaint
has shifted from analysis to supposed clumsiness of style and the
author's coldness toward his characters:

> In *The Princess Casamassima*, Mr. Henry James, without relinquish-
> ing that persistent attention to detail which has militated against the
> success of his later novels, has really written with more breadth and
> verility [sic] than one has come to expect. There are scenes of actual
> distinctness, almost robustness of treatment, not a few of them being,
> common with in [sic] most of the characters, particularly unpleasant.
> The final tragedy is graphic with all the sensationalism of a serial in
> illustrated weeklies of a certain order.

Yet this concessive tone is soon abandoned:

> The book, however, is so intolerably provoking in involved phraseol-
> ogy, its wordiness, everlasting wordiness is so terrific; and the sentences
> and paragraphs are frequently put together in Mr. James's most slip-
> shod, slovenly manner that getting at the force of the story is a con-
> fusing process.

"Slipshod" and "slovenly," applied to James's style, probably express a protest against syntax that is unusually deliberate and complex. A further charge that his "moral philosophy" tends toward "a
profound heartless indifferentism toward life, death and everything
which concerns the human species, bad or good,"[23] is more difficult
to interpret, but "heartless" is no doubt the operative word: James
is on the side of the head, that is, of excessive analysis.

   This charge was repeated in more succinct form by the reviewer
for the solidly middlebrow *Lippincott's:* in *The Princess* James
resembles "an inventor explaining the mechanism by which his
puppets are made to play their appointed parts."[24] And the anonymous reviewer for the relatively highbrow *Critic* grew heated in

developing the charge of excessive analysis. The reader, he asserted, "traverses 134 pages—close, compact, unparagraphed—a snarl of interminable analysis," before the Princess herself enters the story. The critic spoke explicitly of "the failure of this socialistic drama, endlessly delightful as it is to the lover of interpretations, of emotions analytically examined, of hairs radiantly split, of spectroscopic gratings capable of dividing a ray of light into 32,000 lines to the square inch, or of intellectual engines describing 150,000 sensations to the twenty pages."[25] Even so, a few lines later this same reviewer could call the book "an entrancing bundle of emotions and conversations, of eccentric freaks of the analytical imagination."

The ambiguities and apparent self-contradictions in reviews of *The Bostonians* and *The Princess* suggest that James had imposed a strain on the corps of American critics by the genuine novelty that was appearing in his technique. The most perceptive among them intuitively recognized that he was effecting a revolutionary change in the genre but, lacking a theoretical frame within which to place the emergent innovations, they reacted sometimes with bursts of acknowledgment of a strange pleasure in the new effects, and again with anxiety because these effects violated established expectations of novel readers.

During the early 1890s, when James was devoting himself primarily to the London theater, his relations with his American audience remained largely unchanged. He published a few tales in magazines and brought out several volumes collecting these and earlier short fictions. American reviewers continued to complain about "monotonous analysis,"[26] and in 1893 an anonymous reviewer of the collection *The Real Thing* in the *Literary World* reiterated the time-honored charge that James's "imagination vivisects but does not soar."[27] Yet the beginning of a shift in emphasis can perhaps be detected in William Morton Payne's comment about this same volume that "Our enjoyment of his stories . . . must be accounted for almost entirely by the subtlety of his analysis and the polish of his style."[28] And an anonymous reviewer in the *Critic*, while reviving the old distinction that James was writing not fiction but "social studies," said that the tales collected in *The Private Life* in 1893 were executed not only with "literary skill" but also with "all the wonderful insight into human psychology of a tireless explorer in such regions."[29]

It would be forcing the evidence, however, to maintain that critical opinion in general was growing more favorable toward James's analytic method. In a joint review of *The Wheel of Time* and *The Private Life*, Annie R. M. Logan continued to reveal a divided mind about his work. She conceded handsomely that "By the perfection of his rendering of an episode, a situation, a state of mind or soul, he has achieved unique distinction." Yet she declared that the two books under consideration left with her an impression of "form without substance, of fine-spun elusive phantoms with no claim on emotional regard, and rather irritating to the intelligence."[30] Down to the last years of the decade the reviews continued to express similarly mixed attitudes: approval for the writer's immense skill and command of the language along with bewilderment or annoyance at the coldness resulting from his austere maintenance of distance between characters and implied author, and his determination to register the last nuances of his characters' feeling.

Nevertheless, the issues gradually become clarified as reviewers (if not readers in general) gain the ability to see what James is trying to do, or is actually doing, and are able to distinguish this topic from the question of whether they enjoy the result. Thus an anonymous reviewer of *The Wheel of Time* in the *Critic* approached the attitudes that would eventually prevail when he observed that James

is suggestive in his own way; he appeals to the imagination . . . through the friction of mind upon mind rather than through any excitement of incident or of emotion.

The keenness of his analysis of the intricacies of action is a perpetual delight . . . A character appeals to him in proportion as it is complicated . . .[31]

By the end of 1893 Payne, noting that the short stories of James had often been characterized by "a sacrifice of interest to the subtlety of analysis," was able to report that "the more recent of these stories, while losing nothing in subtlety, have distinctly gained in interest."[32]

In the later 1890s there is a perceptible (although still muted) change of tone in the reviews of *The Spoils of Poynton* and *What*

Maisie Knew. Some critics, to be sure, continue the familiar objections to James's fondness for analysis, but we encounter for the first
time an explicit defense of it. An anonymous reviewer for the
Outlook declared that "Mr. James is here as delicately analytical, as
deliberate in manner, as subtle in suggestion, as he has ever
been. . . ." This critic also noted an important but previously
neglected element in James's work in asserting that there is a "keen
humorous sense perceptible throughout. . . ."[33] Even though
critics might not be prepared to endorse the analytical mode in
fiction, they were beginning to recognize that a case might be made
against the long-standing, dogmatic proscription of it. The reviewer
for *Public Opinion* acknowledged (rather ponderously, to be sure)
that James had some precise purpose in mind, that he was "fully
aware of the abatement of interest attending the reading of some of
his work from overwrought analysis and operosity."[34] And an
anonymous critic in the New York *Times* records the dawning
recognition that the stratification of the reading public offered a
reasonable line of development for a writer who was prepared to
sacrifice wide appeal for a mass audience in order to address himself
to a self-conscious elite. James, declared this reviewer, was a new
kind of novelist, "pleasing to one's finer senses":

> He allows nothing for the romantic taste that is in us all. The study
> of character is his single aim, but it is invariably study pursued with
> no idea of giving the shallow entertainment; with no dwelling upon
> eccentric traits humorously, with no tenderness for the weak, with no
> appeal either for laughter or for tears . . .[35]

And the *Book Buyer,* also reviewing *Poynton,* is quite explicit in
noting that the reading public embraces distinct segments having
specialized tastes in fiction. The reviewer, George M. Hyde, declares
flatly that *Poynton,* "throughout . . . is a tale of thrilling psychological interest," and that "the novelist, as by a search-light,
ha[s] swept the remotest nooks of the human soul." But he realizes
that this opinion is not universally shared:

> There seems to be a disposition, in our own country, to debate the
> question whether analysts, any more than the naked savages, "have
> a right to land." Many persons who blunder through life in entire

oblivion of its fascinating subtleties deny that these exist except in the mind of the novelist. The *nuances* which he delights in belong less to the subject, they say, than to his way of viewing it. They decline to delve for motives under whatever auspices. They decry the purely artistic attitude, and, as vaguely, clamor for the picturesque.[36]

One gathers that the critic disagrees, although timidly, with the opponents of the "purely artistic attitude."

Critical discussion of *What Maisie Knew* (1898) raised the final major problem posed by James's work for American readers: the question of morality (particularly, of course, sexual morality). Objections of this kind had been heard as early as the 1880s. Grant Allen had declared that *The Portrait of a Lady*, although a masterpiece in a literary sense, was one of the most "disheartening" books ever written because of its moral emptiness and decadence. Allen asserted that James's influence in general was "morbid and unwholesome."[37] In 1885 a reviewer in the *Literary World* had declared that *Georgina's Reasons* "is simply a study in depravity, as revolting as it could well be."[38] But the morality of James's work did not give rise to extended discussion until the publication of *Maisie*. The *Independent*, despite its explicit religious orientation, was prepared to admit that this novel, "regarded as exotic, strenuously forced and singularly specialized fiction, . . . is worth studying," but declared nevertheless that the book offered "minute analysis of very disagreeable and (to a perfectly healthy mind) uninteresting experiences and aspirations. . . ."[39] The *Bookman* seemed untroubled by the question of morality; blandly merging two traits which in earlier years had seemed irreconcilable, it praised "the wonderful dramatic analysis of this marvelous book."[40] *Public Opinion*, also untroubled by the moral question, praised the author's technical skill—"the manner in which everything is subordinated to the child's mind," a device "so remarkable that it alone would make the work extraordinary."[41] But the *Outlook*, while acknowledging that James maintained "his usual subtlety of analysis," and that the "study" was "truthful," found it "not an agreeable one," and concluded flatly that "the book is tedious."[42]

The review in the relatively highbrow *Critic* suggests that a genuine interest in James's technique might draw attention away from the adulteries Maisie is obliged to ponder:

Four of the chief characters are impossible persons of incredibly abandoned conduct, but by the simple expedient of filtering their complicated and corrupt story through the mind of a child, who apprehends neither morality nor immorality, but only kindness or cruelty, vulgarity or refinement, it is presented to the reader who is wiser than Maisie, as void of offence as it came to her. . . . it is certain that [the personages] move before one flatly and jerkily, like a procession of shadow pictures thrown upon a screen. They are Punch and Judy figures in two dimensions only, not rounded sinners of flesh and blood, but this appearance is only one tribute the more to the marvellous success of the psychological feat Mr. James has performed in inclosing himself in the child's mind and ascertaining her point of view . . .[43]

But the review in the *Nation,* which was dogmatic in condemning the novel on moral grounds, showed that James's audience was not yet divided into distinct factions according to the placement of emphasis on moral or on technical questions:

The device of unfolding a tale, not only without a moral, but without morals, through analysis of impressions made on the mind and character of a child, thrusts the burden of impropriety on the mind of the reader. . . .

It is a sad moment for a writer when he mistakes obscure diction for subtle thought, and Mr. James is confronted with this moment more than once . . .[44]

In *The Awkward Age* (1899) James's late manner is fully developed: it exhibits all the features that have aroused antagonism in some readers and critics from that day to this—the omission of conventional "action"; the avoidance of explicit judgment of characters by the use of a variety of perspectives or various "lights"; the restriction of interest to the relations among a close circle of characters belonging to a leisure class; and on the surface at least, the maintenance of a low emotional pitch. A survey of critical reactions to this book is thus a useful device for exhibiting representative attitudes toward James's work in the last phase of his career. In discussing *The Awkward Age* in the *Sewanee Review,* William P. Trent, founder and editor of the journal, who would soon move to a professorship of English at Columbia, expounded what was probably the prevalent attitude toward James in the academic community and among literate but conservative readers. Noting that "Mr. James's fortes are psychological analysis of character and brilliant management of conversation," Trent placed on record his coolness toward these "prime requisites of successful modern fiction." His review expresses the standard genteel disdain for mathematical and scientific ways of thought, which probably persisted in the South in greater strength than elsewhere:

> . . . if psychological analysis has to be carried to a point of subtlety considerably beyond any attempted by Shakespeare or Balzac, and if conversations and character analysis are the two poles around which the ellipse of modern fiction is to be drawn—we are willing to commend the novels of to-day to the careful attention of students of advanced mathematics, and shall content ourselves hereafter with the simple old novelists who were unsophisticated enough to write straightforward stories.[45]

Yet this set of attitudes was already beginning to seem somewhat out of date. The reviewer of *The Soft Side* for the *Outlook*, in the same year, probably spoke for a considerable and growing minority of readers in approving guardedly of James's analytical method:

> The stories are packed with close observation, with keen analysis, and with that delicate skill which is always at the command of Mr. James. They are touched continually with the exquisite delicacy of style which at his best never fails him; but they are too subtle, too psychological, too analytical, for the purposes of fiction.[46]

Generally speaking, criticism of James during the next decade was to register the growth of the awareness that the Victorian novel, with its emphasis on characterization and plot, was undergoing at James's hands a metamorphosis that would create in effect a new genre. This process was of course one of the main aspects of the liquidation of nineteenth-century culture which would occupy the attention of the Anglo-American world of letters down to our own day.

The transformation of the novel as a genre required a revolution in the expectations with which readers approached it. During the colonial period, and long afterward, the novel had been condemned in this country as morally dangerous; even when reading fiction was not forbidden as a sinful indulgence, it had been scorned as a waste of time. The belief that women were the principal devotees of novels, whether true or not, had fostered the notion that such works made only minimal demands on the rational faculties. On this assumption, critics had kept alive the comfortable simplification that novels appeal to the heart, not the head.[47] Much of the objection to James's devotion to analysis of the mental processes of his characters was predicated on a sense that it was inappropriate, a breach of an implied contract with the reader. In making novel-reading into a demanding intellectual exercise, James created a highbrow level of taste distinct from the established middlebrow attitude, the esthetic of the common reader. An anonymous reviewer of *The Sacred Fount* in *Harper's Monthly* (not Howells) spoke for many others in declaring that James

> turns upon his course describing an ellipse, and ellipses within that ellipse—always a faithful following of the psychological involutions in the author's subjective analysis—until the reader of average intelligence is lost in the bewildering maze.[48]

Montgomery Schuyler (best known as a critic of architecture but also a former editor of *Harper's Weekly* and an editor of the New York *Times*) commented in 1902 on the changed demands made on readers by James's late writing:

> . . . nothing could more unfit a reader for the intent attitude of mind proper to studying his psychology than the languid perusal which is all that most contemporary fiction requires or repays. And this professor of psychology is increasingly indifferent to the comfort of his students. It is perhaps mainly in consequence of this indifference that the first living master of psychological fiction, at least in the English language . . . finds himself compelled to flit from publisher to publisher. . . .
> The subtleties of analysis which the novelist goes into might do if there were a "public" of mighty poets and subtle-souled psychologists. But when the plain man finds what looks to him like an attempt to decompose immediate intuitions and trace their steps, when by hy-

pothesis they haven't any, it is no wonder the plain man revolts. Whether it is "worth going through so much to learn so little" is a question that each reader must and will answer for himself . . .[49]

Yet the *Independent,* surprisingly enough, was inclined to be more judicious in its assessment of James's procedure in *The Wings of the Dove.* Acknowledging that much of the novel seems to consist of "little else but long, dull paragraphs of emotional tergiversation, wherein one loses all sense of direction for lack of one little clue, one single clear straightforward word. . . ," the reviewer declared that James's "intention is perfectly plain." He writes as he does "in the hope of catching those subtle 'psychic' states which he reproaches Flaubert with having neglected."[50]

A note of irritation, and the notion that James is accessible only to a small coterie of readers, continued to be more common. Montgomery Schuyler declared in 1903 that "Mr. James is a writer's writer. The world of his psychology, a world of theorems and problems, is as far as may be from the world in which the common novel-reader aspires to lose himself."[51] And a reviewer for *Current Literature* repeated the allusion to mathematics, declaring that James's fictive world is one

> in which nothing ever happens and in which the characters are not distinguished by any action save a subtle psychological mentality so beloved by Mr. James and so detested by the average reader who cannot follow the deceptive quality of the author's style. Delicate and refined analysis of character marks this book, but it is analysis raised to the Nth root [sic]. The majority of readers are unacquainted with the binomial theorem as employed by Mr. James.[52]

To quote only one more contemporary statement about James's audience, a reviewer of *Julia Bride* maintained in 1909 that "The average busy man or woman of today will hardly read a novel of this sort with enthusiasm. It is too much of a psychological puzzle to prove either instructive or entertaining."[53]

Nevertheless, amid the balanced and tentative opinions of James's work characteristic of the American world of letters during the early years of the twentieth century, there appeared several enthusiastic devotees—most of them women, as it happens, and

none of them associated with a university. One of the earliest of the
self-proclaimed Jacobites, Cornelia Atwood Pratt, was virtually
alone in responding with enthusiasm to *The Awkward Age*. (She
declared it to be "ahead of anything he [James] has yet produced, for
subtlety and acute insight."[54]) A second consistent admirer was
Carolyn Shipman, who declared in the *Book Buyer* in a notice of
The Sacred Fount that "Whatever may be said of Mr. James's
increasing mannerisms of style, no one who is interested in human
nature can fail to care for what he has to say."[55] Perhaps the most
significant indication of a basic shift in critical assumptions was the
willingness of such readers to make James's analytical method—
even when it is called "dissection"—an element in his excellence
rather than a weakness to be apologized for.

> In "The Ambassadors" [wrote an anonymous reviewer in *Current
> Literature*] . . . James has carried the development of his later
> characteristics to a finer issue than in any of his other well-known
> works. The laboriously subtle analysis of character; the minute dis-
> section, so to speak, of every nerve; the painfully precise description
> of every step of thought and action, with frequent sudden breakings-
> off of articulate speech, making silence all the more suggestive—all
> are found in this volume . . .[56]

Yet another female partisan of James, Elizabeth Luther Cary, declared in *Scribner's* in 1904 that "nowhere else can we learn the
jealously guarded secrets of the mind and heart that become important to us in proportion as our fellow beings render them inaccessible. In the novels of Henry James we drink deep of this peculiar
satisfaction."[57]

Even when such critics acknowledged difficulties in James's
later manner, they were capable of making observations that stand
up well three-quarters of a century later. An anonymous reviewer for
the *Outlook*, for example, asserted in 1904 that the style James
employed in *The Golden Bowl* was "charged . . . with the fullest
subtleties of the literary art, mastered by infinite patience and
practiced with consummate skill—a skill compounded of rare psychological insight and of extraordinary feeling for the value of
words." When this critic adds that "'The Golden Bowl' is a marvel
of adroitness . . . ," he has established a sufficiently solid recog-

nition of James's value to be able to add, without doing an injustice to the novel, that

> As a piece of fiction, as a work of literature pure and simple, it has grave defects; it is too circuitous, too heavily laden with description, comment, and suggestion; the analysis is carried so far and pressed so hard that the current of the story is hindered and the larger lines are lost in a mass of detail.[58]

One or two of these reviewers had intimations of the important truth that James's late work pointed in the direction fiction would take in the coming decades, the era we have learned to call "modernist." For even before realism in fiction had completely supplanted the conventions generated by idealism, many leading European writers had begun to recognize the limitations of realism. Theoretically, of course, the realistic program of psychological analysis, of telling the truth about the passions and motives of men and women as well as about their actions (as set forth for example in Howells's "Criticism and Fiction"[59]), was simple and unambiguous to the point of seeming self-evident. But the positivism, the absolute disjunction between the outer world of hard fact and the inner world of mental process that was the philosophical basis for the theory of realism, could not support equal emphasis in fiction on the outer and inner realms: the outer realm of observable, even measurable facts seemed undeniably more real than an inner realm that embraced fantasies and delusions along with perceptions and inferences. Howells, poorly equipped as he was for philosophical thought, was at least vaguely aware of this difficulty when he declared that he analyzed as little as possible.[60]

James's method moved in the opposite direction, leading eventually to a psychological conception of reality. Elizabeth Luther Cary, hampered to be sure by an archaic vocabulary, noted with great acuteness in 1905 that *The Golden Bowl* marked an advance beyond Balzac; for his "all-encompassing love of reality took insufficient account of the pressure of complex circumstances upon the individual soul. His divination stopped, in fact, at the door of the soul." "Soul" is not a term with which anyone could do justice to the psychological and ontological discoveries that were in process around her. But she was looking in exactly the right direction when

she declared that "With the novels of Henry James there is excitement for the adventurous mind of entering this new field, these
undiscovered regions of reality."[61]

Mrs. Logan was able to develop this insight further by probing
more deeply into James's narrative technique. She said that *The
Ambassadors* "suggests the substance of what seems to be Mr.
James's theory of realistic representation. . . ," to wit:

> His plan of campaign is to show how an incident or train of incidents
> appears to one or two or twenty people, what each makes out of the
> complication, and how his interpretation affects his feeling, judgment,
> conduct. . . . By delineating interested observers thus pleasantly
> engrossed in an interesting situation, Mr. James tries for very extended
> realistic representation, and very difficult, because part of his task is,
> through his various readings of the vain appearance, to show the truth,
> or, at least, to indicate the greater probability.[62]

With the phrase "indicate the greater probability" Mrs. Logan
abandons nineteenth-century positivism and places her foot on the
threshold of the philosophic discoveries underlying twentieth-
century fiction.

H. G. Dwight made a similar point, more elaborately, in a
perceptive essay published in *Putnam's* in 1907:

> Different as he [James] is from Ibsen, from Maeterlinck, from D'An-
> nunzio, from Hauptmann and Sudermann, from Paul Bourget and
> Anatole France and the Russians [wrote Dwight], he is yet one with
> them, as they are one with each other, in a certain unmistakable trend
> of modern literature. . . . The repression of action in his later
> novels, the tracing of action to its secret sources, which to a public
> schooled in the older tradition seems perverted or ridiculous, may be
> primarily a matter of constitution; but it has the closest possible
> relation to a movement in the wider world of letters. If there is any-
> thing at all in what we vaguely call the *Zeitgeist*, it would seem that
> as consciousness increases, as we become more trained to the conse-
> quence of much that we have regarded as inconsequent, books like
> "What Maisie Knew" and "The Sacred Fount" and "The Golden
> Bowl" will take on for us a new significance.[63]

Although this is hardly the place to undertake a survey of
twentieth-century European and American fiction, I am tempted to

quote a few sentences from a recent article on James's theory of
the novel by James E. Miller, Jr. Miller points out that James maintained a complex and subtle notion of representation: ". . . life or
reality is never statically 'out there' but always shaped, given significance by the vision of the perceiver." And again: "By the time we
have moved from the notion of art representing life to the notion of
art making life, we have moved from the nineteenth-century realistic
novel to the twentieth-century 'new novel,' from Gustave Flaubert
or George Eliot or William Dean Howells to Vladimir Nabokov or
Alain Robbe-Grillet or Gabriel García Márquez." James's assertion
that "It is art that makes life" is consonant with Robbe-Grillet's
contention that the style of the novel "*constitutes* reality," and the
declaration of Carlos Fuentes that the death of bourgeois realism
"announces the advent of a literary reality much more powerful."[64]

James's position in the last decade of his life was paradoxical. He
was widely recognized as an important writer, yet his books did not
sell. He had in consequence formed the low opinion of both readers
and critics that is recorded in such stories as "The Death of the
Lion" and "The Next Time." He was not unique in his antagonism
to the prevalent taste in literature and the popular culture it expressed, for scorn of the bourgeoisie had long been almost routine
among men of letters in France, had penetrated English society in
the Aesthetic Movement of the 1880s, and was widespread in various
European countries on the eve of the First World War. Indeed,
Renato Poggioli observes in *The Theory of the Avant-Garde* that
"the chief function" of modern art "is to react against bourgeois
taste."[65] Many writers at the turn of the twentieth century were
suffering from the malady of alienation that Poggioli describes:
". . . the feeling of uselessness and isolation of a person who
realizes that he is now totally estranged from a society which has lost
its sense of the human condition and its own historical mission" (p.
109). Yet in this regard James resists classification. He was anything
but alienated from the high culture cherished by his friends and
associates, and the elite minority to which they belonged. He
remained loyal to the set of values that he and they called "civili-
sation." In his usage the term referred not only to the heritage of art

absolutely .

received from the past—architecture, sculpture, painting, music, as well as literature—but also (as he made plain, for example, in his commencement lecture "The Question of Our Speech" at Vassar College in 1904) good manners and cultivated intonation.[66] He was not in the least attracted by the bohemian and politically subversive movements such as Dadaism and the like that had begun to appear at intervals (mainly of course on the Continent) during his later years.

James's position was in fact far from simple. The unconventional education provided for him and his siblings by their remarkable father prevented him from receiving the religious indoctrination that was almost universal in Americans of his class. His decision to live abroad showed how free he was of that other basic ingredient of the dominant American ideology, its perfervid nationalism.[67] Although under the pressure of the Civil War the young James could become temporarily infected with the mass emotion and belief, to the extent of declaring that the War had revealed "the clear spiritual insight . . . of great popular impulses" and that the victory of the North had been arranged by God,[68] he almost never thereafter made any allusion to religious ideas.[69] By the time he was twenty-four he was prepared to contradict James Froude's proclamation of the great lesson taught by "history," namely, that "the world is built somehow on moral foundations; that in the long run it is well with the good; in the long run it is ill with the wicked." James's comment was: "If there is one thing that history does not teach, it seems to us, it is just this very lesson. What strikes an attentive student of the past is the indifference of events to man's moral worth or worthlessness."[70]

Such an attitude flouted American popular culture even more flagrantly than it did British. I quoted earlier Santayana's assertion that at the end of the nineteenth century, American religious teachers and philosophers felt the world "was a safe place, watched over by a kindly God, who exacted nothing but cheerfulness and good-will from his children. . . ."[71] Howells was basically in agreement. In 1894 he recalled his delight as a boy in reading Dickens's novels,

. . . where the right came out best, as I believe it will do yet in this world. . . . In that world of his, in the ideal world, to which the real world must finally conform itself . . . a Providence . . . governed all things to a good end . . . Of course it was in a way all crude enough, and was already contradicted by experience in the small sphere of my own being; but nevertheless it was true with the truth which is at the bottom of things. . . .[72]

It might be maintained, in fact, that one of the principal factors accounting for the marked differences between the work of James and that of Howells, especially their later work, was Howells's resolute clinging to this faith that James had altogether abandoned. Whereas Howells's parochial fidelity to the cosmic optimism of American popular culture anchored him in the nineteenth century, James's secular attitude placed him closer to the mainstream of Western high culture. He thus suffered from no inner conflict as his technical development led him toward the creation of a fictional mode appropriate to the skeptical temper of the new century.

Nevertheless, the shock of 1914 jolted James into awareness that he cherished an underlying faith not wholly different from Howells's. James discovered, after all, that he too believed in civilization, in progress, in a moral order in the universe, at a time when writers of the next generation—James Joyce and T. S. Eliot and Ezra Pound—were beginning to perceive Western civilization as an old bitch gone in the teeth, or a Waste Land, and even Mark Twain was declaring (in private) that a review of the history of mankind showed the whole universe to be "a grotesque and foolish dream."[73] Looking back in the bitterness of his eventual disillusionment, James recalled with a kind of wonder "the American, or at least the Northern, state of mind and of life that began to develop just after the Civil War. . . ." "We really," he wrote in 1915, "we nobly, we insanely (as it can only now strike us) held ourselves comfortably clear of the worst horror that in the past had attended the life of nations." It seemed then that in four years of carnage the United States "had exactly *shed* the bad possibilities, were publicly purged of the dreadful disease [of war] which had come within an inch of being fatal to us, and were by that token warranted sound forever, superlatively safe. . . ." This deluded optimism, he continued, had spread "over a considerable part of the earth." Even the defeat of Austria by Prussia and the German victory in the Franco-Prussian war could seem steps on the way to peace because they consummated the unification of Germany. "So it was," James concluded, "that our faith was confirmed—violence sitting down again with averted face, and the conquests we felt the truly golden ones spreading and spreading behind its back."[74]

A few months before the outbreak of the War, James had written to Henry Adams a letter (often quoted in later years) that contained a remarkable declaration of faith:

> You see I still, in presence of life. . . , have reactions—as many as possible—and the book I sent you [*Notes of a Son and Brother*] is a proof of them. It's, I suppose, because I am that queer monster, the artist, an obstinate finality, an inexhaustible sensibility. Hence the reactions—appearances, memories, many things, go on playing upon it with consequences that I note and "enjoy" (grim word!) noting. It all takes doing—and I *do*. I believe I shall do it yet again—it is still an act of life.[75]

This is a different set of ideas from those embodied in James's references to progress and civilization, emblems of a faith which he shared with spokesmen for the popular culture. The notion of the autonomous artist, dependent exclusively on his own sensibility and reactions, was a highbrow notion then and continues to be so now, although its practical implications for the development of literature and the other arts have been much more fully worked out.

Would James's conception of the obstinate finality of the artist have enabled him to recover from the shock of the War? We shall never know, because the few months of life left to him were taken up with frenzied exertions in behalf of refugees and wounded soldiers, or any other task he could find that aided the cause of the Allies. But we do know he was desolated by the revelation that he and everyone he knew had been so badly mistaken in their conception of reality during the confident decades of the later Victorian era. As he stated the problem to Hugh Walpole shortly before his last illness, "the subject-matter of one's effort" had after all been a supposed reality outside the artist's mind, and the War had shown that reality to be mere deceptive appearance:

> I have in a manner got back to work—after a black interregnum; and find it a refuge and a prop—but the conditions make it difficult, exceedingly, almost insuperably, *I* find, in a sense far other than the

mere distressing and depressing. The subject-matter of one's effort has become *itself* utterly treacherous and false—its relation to reality utterly given away and smashed. Reality is a world that was to be capable of *this*—and how represent that horrific capability, *historically* latent, historically ahead of it? How on the other hand *not* represent it either—without putting into play mere fiddlesticks?[76]

It may be that James, like Mark Twain, would have found his imagination paralyzed when he fully assimilated twentieth-century reality.

modern history has
played havoc with the
liberal faith and dulled
~~underminded it~~
one of its shinier and
most serviceable
ornaments ~ the
american historical
tradition,

Notes

Chapter 1   The Issues

1. F. O. Matthiessen, *American Renaissance. Art and Expression in the Age of Emerson and Whitman*, New York, 1941, pp. viii–xi.
2. Matthiessen's work has recently been analyzed at length in Giles B. Gunn's sympathetic study, *F. O. Matthiessen: The Critical Achievement*, Seattle, 1975. On paradoxes and apparent contradictions in Matthiessen's thought, see esp. pp. xix–xxi, 13–17.
3. Nathaniel Hawthorne, "Preface," *The Marble Faun: or, The Romance of Monte Beni* (Centenary Edition), eds. William Charvat et al., Columbus, Ohio, 1968, p. 3.
4. "The Genteel Tradition in American Philosophy," in *The Genteel Tradition. Nine Essays by George Santayana*, ed. Douglas L. Wilson, Cambridge, Mass., 1967, pp. 38–64.
5. Perry Miller, "The Romance and the Novel," in *Nature's Nation*, Cambridge, Mass., 1967, pp. 255–256.
6. The circumstances of Hawthorne's letter are discussed more fully in my article, "The Scribbling Women and the Cosmic Success Story," *Critical Inquiry*, 1 (September 1974), 47–70.
7. Ibid., 1:48–50.
8. William Charvat, *The Origins of American Critical Thought: 1810–1835*, Phila., 1936, pp. 4–6. —Lewis P. Simpson has described the ideal of a Republic of Letters presided over by an American clerisy in *The Man of Letters in New England and the South. Essays in the History of the Literary Vocation in America* (Baton Rouge, 1973).

9. Ann Douglas, in *The Feminization of American Culture* (New York, 1977), has recently investigated at considerable length the role of women in this transformation. Chapter 7 of her work, "The Periodical Press," is particularly informative concerning the changes in the reading public during the middle decades of the century. A casual remark of Ms. Douglas states one of the principal theses of my own study. In the opposition of Thoreau, Cooper, Melville, and Whitman to feminism, she says, "we see the beginnings of the split between elite and mass cultures so familiar today" (p. 6). Looking back from the 1890s, she declares: "Certain forms of deprivation and exclusion had made middle-class American women, the readers and consumers of the nation, and the men who imitated, flattered, and exploited them, logical heirs to the anti-intellectual tradition in American culture; and they had conspired willy-nilly with changing historical circumstances to make anti-intellectualism the tradition in American culture" (p. 328).

10. Although I should modify Richard Chase's definitions slightly, I am glad to be able to invoke his example as a precedent for using these terms (*The American Novel and Its Tradition*, New York, 1957, pp. 9–11). —According to the *OED Supplement*, the adjective "highbrow" came into use in the 1880s as a back formation from "highbrowed." This word, in turn, was clearly borrowed from phrenology. Poe, for example, declared that William Cullen Bryant's forehead was "broad, with prominent organs of Ideality" (quoted from *The Literati*, New York, 1850, in John D. Davies, *Phrenology Fad and Science. A 19th-Century American Crusade*, New Haven, 1955, p. 121). The *OED Supplement* records "lowbrowed" as an adjective as early as 1855, but "lowbrow" as substantive and adjective only after 1900. The wide currency of "highbrow" and "lowbrow" dates from Van Wyck Brooks's use of the words in *America's Coming-of-Age* (New York, 1915). I know of no occurrence of "middlebrow" earlier than Virginia Woolf's essay of that title, in *The Death of the Moth* (1942), reprinted in *Collected Essays* (ed. Leonard Woolf, 4 vols., London, 1966–67, 2:199, 202–203). The best known recent discussions of the topic are Russell Lynes's "Highbrow, Lowbrow, Middlebrow," originally in *Harper's* (February 1949), reprinted with slight revisions in *The Tastemakers* (New York, 1954, pp. 310–333); and Dwight Macdonald's "Masscult and Midcult," originally in *Partisan Review* (1960), reprinted in *Against the American Grain. Essays on the Effects of Mass Culture* (New York, 1962, pp. 3–75). Lynes used the terms loosely and satirically. Macdonald was more serious but he too failed to press toward clear definitions. One may hope that the recent increase in scholarly concern with popular culture will

result in a more systematic analysis of the stratification of American literary taste.

11. Louis James, *Fiction for the Working Man, 1830–1850. A Study of the Literature Produced for the Working Classes in Early Victorian Urban England*, London, 1963. James assumes that reading matter produced especially for this audience can be identified by its cheapness. —I have traced the exploitation of Wild Western adventure, introduced into fiction by Cooper in his Leatherstocking series (1823–1842), in *Virgin Land. The American West as Symbol and Myth*, Cambridge, Mass., 1950, repr. 1970, pp. 51–120.

12. Nathaniel Hawthorne, *Mosses from an Old Manse*, in *Complete Works*, ed. George P. Lathrop, 13 vols., Boston, 1897–1898, 2:107. Poe's favorable review of *Twice-Told Tales* in 1842 refers several times to the fact that Hawthorne's work has had only limited sales (reprinted in *Selected Writings*, ed. Edward H. Davidson, Boston, 1956, pp. 440–441).

13. William Charvat, "Introduction," *The Scarlet Letter* (Centenary Edition), Columbus, Ohio, 1962, p. xxii.

14. *The Portable Melville*, ed. Jay Leyda, New York, 1952, pp. 407–409, 417–418.

15. Nathaniel Hawthorne, *Scarlet Letter*, p. 3. James H. McIntosh provides fuller information about this topic in "Hawthorne's Search for a Wider Public and a Select Society," *Forum* (University of Houston), 13 (Winter 1976), 4–7.

16. William Charvat, "Melville and the Common Reader," *Studies in Bibliography* (Papers of the Bibliographical Society of the University of Virginia), 12 (1959), 45, 51–52.

17. Ibid., 12:54–55.

18. *Blackwood's Edinburgh Magazine*, 77 (May 1855), 562–565.

19. Terence Martin, *The Instructed Vision*, Bloomington, Ind., 1961, esp. pp. 136–138.

20. George Santayana, "The Moral Background," in *Character and Opinion in the United States* (1920), reprinted in *The Genteel Tradition*, p. 86. This set of ideas has proved remarkably durable. It is a version of what Henry James would call the great mistake of his generation, exposed traumatically by the outbreak of the First World War (see pp. 164–165). But Saul Bellow's Moses Herzog notes that as late as the 1960s, "bourgeois" Americans continued to believe "the universe was made for our safe use and to give us comfort, ease, and support. Light travels at a quarter of a million miles per second so that we can see to comb our hair or to read in the paper that ham hocks are cheaper than yesterday" (*Herzog*, New York, 1961, p. 50).

21. Rush Welter, *The Mind of America, 1820–1860*, New York, 1975, esp. pp. 3–13, 256–264.
22. David Levin, *History as Romantic Art: Bancroft, Prescott, Motley, and Parkman*, Stanford, Cal., 1959, esp. pp. 26–27.
23. Welter, *The Mind of America*, pp. 143, 147–149.
24. Idealism in esthetics was often called "ideality" (Davies, *Phrenology Fad and Science*, esp. Chap. 10, pp. 118–125).
25. Nathaniel Hawthorne, "Preface," *The House of the Seven Gables* (Centenary Edition), eds. William Charvat et al., Columbus, Ohio, 1965, p. 1.
26. *Blackwood's*, 77:562–565.
27. William Charvat, "The People's Patronage," in *The Profession of Authorship in America, 1800–1870*, ed. Matthew J. Bruccoli, Columbus, Ohio, 1968, p. 249.
28. Emma D. E. N. Southworth, *Retribution; or, The Vale of Shadows. A Tale of Passion*, New York, 1849, p. 108; seen on microfilm.

Chapter 2   Hawthorne: The Politics of Romance

1. Nathaniel Hawthorne, *Scarlet Letter*, pp. xix, 14, 41.
2. Perry Miller, "Jonathan Edwards to Emerson," *New England Quarterly*, 13 (December 1940), 612, 610, 606.
3. Perry Miller has a good brief description of this conflict in the introduction to his anthology *The Transcendentalists*, Cambridge, Mass., 1950, pp. 3–15.
4. Bertha Faust, *Hawthorne's Contemporaneous Reputation*, Philadelphia, 1939, pp. 55, 59; Hans-Joachim Lang, "*The Blithedale Romance:* A History of Ideas Approach," in *Literatur und Sprache der Vereinigten Staaten: Aufsätze zur Ehren von Hans Galinsky*, eds. Hans Helmcke et al., Heidelberg, 1969, pp. 89–90.
5. Nathaniel Hawthorne, *The Blithedale Romance and Fanshawe* (Centenary Edition), eds. William Charvat et al., Columbus, Ohio, 1964, p. 141.
6. *Hawthorne's Reputation*, p. 50.
7. Lawrence S. Hall, *Hawthorne: Critic of Society* (Yale University Studies in English, 99), New Haven, Conn., 1944, pp. 34–35, 43.
8. Charvat, *Origins of American Critical Thought*, pp. 1, 6.
9. Ibid., pp. 12, 20.
10. *North American Review*, 45 (July 1837), 206.
11. Francis Wayland, for example—perhaps the most influential of the pre-Civil War moral philosophers—asserted that "the precepts of religion"

command us "to obey magistrates" and to submit "to every ordinance of man for the Lord's sake" (Wilson Smith, *Professors and Public Ethics: Studies of Northern Moral Philosophers before the Civil War*, Ithaca, N.Y., 1956, p. 135).

12. *North American Review*, 64 (April 1847), 414.

13. Although Cooper was in general hostile to both the radical and the conservative varieties of New England thought, he was thoroughly in accord with the common-sense metaphysical and epistemological assumptions that Bowen expounds in his criticism of Emerson. Cooper's views are clearly set forth by H. Daniel Peck in his recent study, *A World by Itself: The Pastoral Moment in Cooper's Fiction*, New Haven, Conn., 1977, pp. 9–17.

14. Hawthorne, "Preface", *The House of the Seven Gables*, p. 1.

15. *Nature* (1836), reprinted in *Selections from Ralph Waldo Emerson* (Riverside Edition), ed. Stephen E. Whicher, Boston, 1957, p. 36.

16. Nathaniel Hawthorne, "Preface", *The Snow-Image, and Other Twice-Told Tales*, in *Complete Works*, 3:386.

17. Hawthorne, "Preface", *The Marble Faun*, p. 3.

18. Quoted in Faust, *Hawthorne's Reputation*, p. 104.

19. Review of *The Scarlet Letter* by Anne W. Abbott in *North American Review*, 71 (July 1850), quoted in *Hawthorne's Reputation*, p. 75; review of *Blithedale* by Charles Francis Adams, Jr., in *Harvard Magazine* (1855), quoted in Lang, *"The Blithedale Romance:* A History of Ideas Approach," p. 90. —It should be added that the *North American Review* was not bigoted in its hostility toward Hawthorne. An unsigned review of *The House of the Seven Gables* and *The Blithedale Romance*, published in January 1853 (76:229), declares: "His golden touch . . . imposes no superficial glitter, but brings out upon the surface, and concentrates in luminous points, the interior gilding, which is attached to the meanest objects and the lowliest scenes by their contact with the realm of sentiment, emotion, and spiritual life." The review was probably written by Andrew Preston Peabody, who became the editor of the *Review* in 1853.

20. *North American Review*, 71 (July 1850), 139–140; John G. Cawelti, *Adventure, Mystery, and Romance*, Chicago, 1976, p. 26.

21. Charles Feidelson, Jr., *Symbolism and American Literature*, Chicago, 1953, p. 11.

22. Hawthorne, *Complete Works*, 2:381, 401–402, 406.

23. Frederick Crews, *The Sins of the Fathers: Hawthorne's Psychological Themes*, New York, 1966, Chap. 5, esp. pp. 86–87.

24. I have made some use here of Jack Kligerman's article, "A Stylistic

Approach to Hawthorne's 'Roger Malvin's Burial,'" *Language and Style,* 4 (Summer 1971), 188–194.

25. Emerson, *Selections,* ed. Whicher, p. 78.

26. Leo Spitzer, "Linguistic Perspectivism in the *Don Quijote,*" in *Linguistics and Literary History: Essays in Stylistics,* Princeton, N.J., 1948, pp. 41–85, esp. 50, 61, 68–73. —One must keep in mind, however, that Hawthorne was often troubled by the lack of "materiality" in his fiction; his notebooks, through the years, show him setting down for use in stories descriptions of "actual places, incidents, and people drawn from his own observation and experience" (Randall Stewart, "Introduction," in *The American Notebooks,* New Haven, Conn., 1932, p. xxvii).

27. Emerson, *Selections,* ed. Whicher, p. 31.

28. Michael D. Bell, *Hawthorne and the Historical Romance of New England,* Princeton, N.J., 1971, p. 142. —Edwin Fussell takes a similar view of the sermon, calling it "surprisingly political," and noting that it "bears . . . a curious resemblance to the demagogic oratory of Hawthorne's day. . . ." (*Frontier: American Literature and the American West,* Princeton, N.J., 1965, p. 112). But Sacvan Bercovitch interprets the sermon differently, saying that Dimmesdale's "message here unmistakably sums up the author's hope for his country . . ." ("Horologicals to Chronometricals," *Literary Monographs* (Madison, Wis.), 3 (1970), 123).

29. The widespread commitment to the notion of Manifest Destiny in the period 1830 to 1860 is documented fully by Wilson Smith in *Professors and Public Ethics,* esp. pp. 17, 78.

30. James Joyce, *A Portrait of the Artist as a Young Man,* New York, 1964, p. 215.

31. Jean-Paul Sartre insists on the obligation of the novelist to maintain the freedom and therefore the unpredictability of his characters in his celebrated attack on François Mauriac ("François Mauriac and Freedom," in *Literary Essays,* trans. Annette Michelson, New York, 1957, pp. 7–23, esp. p. 23). In another well-known essay Sartre declares: "We want to drive providence from our works as we have from our world" ("The Situation of the Writer in 1947," in *What Is Literature?* trans. Bernard Frechtman (Colophon Books), New York, 1965, p. 223).

32. Spitzer notes that more recent artists and thinkers such as Gide, Proust, Conrad, Joyce, Virginia Woolf, and Pirandello, in imitating Cervantes' perspectivism, have not accepted the presumption of unity behind it, so that "in their hands, the personality of the [implied] author is allowed to disintegrate" (*Linguistics and Literary History,* p. 72). Cervantes, on

the other hand, takes for granted "the old Neo-Platonic belief in an artistic Maker who is enthroned above the manifold facets and perspectives of the world" (p. 61). His untroubled Christian orthodoxy appears in the fact that "High above this world-wide cosmos of his making, in which hundreds of characters, situations, vistas, themes, plots and subplots are merged, Cervantes' artistic self is enthroned, an all-embracing creative self, Naturelike, Godlike, almighty, all-wise, all-good— and benign   . . ." (pp. 72–73). Hawthorne is much less secure in his faith in a benign and omnipotent God, but he has not lost this faith altogether.

33. Notice, however, that Hawthorne's narrative method gives rise to the conflicting interpretations of Dimmesdale's sermon we have just noted.

34. Cf. Wayne Booth, *The Rhetoric of Fiction,* Chicago, 1961, pp. 158–159.

35. [Salem, 4 October 1840], in *Love Letters of Nathaniel Hawthorne,* ed. Roswell Field, 2 vols., Chicago, 1907, 1:225.

36. Ernest Tuveson, *Imagination as a Means of Grace: Locke and the Aesthetics of Romanticism,* Berkeley, 1960.

Chapter 3   Madness of Ahab

1. Melville to Hawthorne, Pittsfield, Mass., 16 April 1851; "Hawthorne and His Mosses," quoted from *Literary World,* 17 August 1850, reprinted in *Moby-Dick: An Authoritative Text, Reviews and Letters by Melville, Analogues and Sources, Criticism* (Norton Edition), eds. Harrison Hayford and Hershel Parker, New York, 1967, pp. 555, 541–542. Subsequent citations to this edition appear in parentheses in the text. It may be significant that Melville associated madness with the dark truth before he conceived the character of Ahab.

2. T. S. Eliot, "Hamlet and His Problems," in *Selected Essays 1917– 1932,* New York, 1932, pp. 124–126.

3. Herman Melville, *Mardi and a Voyage Thither* (Northwestern-Newberry Edition), eds. Harrison Hayford, Hershel Parker, and G. Thomas Tanselle, Evanston, Ill., 1970, pp. 419, 605–609.

4. Herman Melville, *Redburn: His First Voyage* (Northwestern-Newberry Edition), eds. Harrison Hayford, Hershel Parker, and G. Thomas Tanselle, Evanston, Ill., 1969, p. 61.

5. Isaac Ray, *A Treatise on the Medical Jurisprudence of Insanity* (1838; reprinted in John Harvard Library), ed. Winfred Overholser, Cambridge, Mass., 1962, pp. 122–123; James C. Prichard, *A Treatise on Insanity*

and Other Disorders Affecting the Mind (1835; reprinted Philadelphia, 1837), p. 30.

6. The term "moral insanity" was introduced by Prichard as a translation of *"manie sans délire,"* a phrase which had apparently first appeared in 1801 in Philippe Pinel's *Traité médico-philosophique sur l'aliénation mentale, ou la manie* (Norman Dain, *Concepts of Insanity in the United States, 1789–1865*, New Brunswick, N.J., 1964, pp. 73–74). —Although I was not aware of it when I wrote the version of this chapter that appeared in the *Yale Review* in 1976 (66, Autumn, 14–32), in 1972 Armin Staats had drawn upon Prichard's *Treatise* to support his thesis that the character of Ishmael has a structural relation to that of Ahab based on a taxonomy of mental disorders: ". . . Ahab, the monomaniac, is conceived as the second phase of a psychopathological process, of which the melancholic first-person narrator Ishmael represents the introductory phase . . ." ("Melville—*Moby-Dick*," in *Der amerikanische Roman von den Anfängen bis zur Gegenwart*, ed. Hans-Joachim Lang, Düsseldorf, 1972, p. 123; my translation). Staats also cites Esquirol, but he is not interested in the distinction between disturbance of the emotions and disturbance of cognition. On the contrary, he maintains that in Prichard's system Ishmael's "melancholy," a neurotic condition, as well as Ahab's "monomania," a psychotic condition, involves derangement both of feeling and of perception or cognition (pp. 124–125). He recognizes that Ishmael recovers from the almost hypnotic spell Ahab casts on the crew (a process depicted in "The Try-Works," Chapter 96, p. 140); and this means that Ishmael attains a rational perspective on Ahab's quest. Ahab, on the other hand, once he falls a victim to monomania, is a static character, undergoing no change. Ahab's malady means that he carries to his death not only a diseased emotion (his uncontrollable determination to avenge himself on his adversary) but also a distorted perception or *idée fixe* (his paranoid conception of the whale as the symbol of some evil supernatural power). —In my opinion, Staats's interpretation, while admirably suggestive and by no means entirely erroneous, is over-schematic. He calls both Ishmael and Ahab "ideological characters" (p. 120), and to some extent they undoubtedly are. But Staats makes them conform to an ideology that with its Marxist and Freudian aspects seems rather Staats's than Melville's, and the ideological view leads him to disregard contradictions and gaps in the book as it was actually written.

7. "Commonwealth vs. Abner Rogers, Jr.," *Reports of Cases Argued and Determined in the Supreme Judicial Court of Massachusetts*, ed. Theron

Metcalf, vol. 48, Boston, 1851, p. 502. In the late 1840s Melville was a frequent visitor in the home of Judge Shaw, who owned a large and varied library (Merton M. Sealts, Jr., *Melville's Reading: A Check-List of Books Owned and Borrowed*, Madison, Wis., 1966, p. 16). But there is no evidence that Melville read this opinion.

8. Ray, *Medical Jurisprudence*, Appendix I, pp. 343–350.

9. "Definition of Insanity—Nature of the Disease," *American Journal of Insanity*, 1 (October 1844), 109. Brigham was editor of the journal. —Thomas C. Upham, Professor of Mental and Moral Philosophy in Bowdoin College, makes a similar point about resistance to the notion of moral insanity in *Outlines of Imperfect and Disordered Mental Action* (New York, 1840, pp. 57–59). Upham's book was the one hundreth volume in Harper's Family Library. Many volumes in this series were in the ship's library of the man-of-war *United States* while Melville served in her in 1843–44 (Sealts, *Melville's Reading*, p. 127, note 44). But again, there is no evidence that the writer read either of these discussions of moral insanity.

10. Melville, *Moby-Dick* (Norton Edition), pp. 590–601; Leon Howard, *Herman Melville: A Biography*, Berkeley (1951), reprinted 1958, p. 184.

11. Ray, *Medical Jurisprudence*, pp. 109, 112.

12. Max Byrd, *Visits to Bedlam: Madness and Literature in the Eighteenth Century*, Columbia, S.C., 1974, pp. 106, 117, 118.

13. Staats, "Melville—*Moby-Dick*," p. 112.

14. Brigham collects such material in "Insanity—Illustrated by Histories of Distinguished Men, and by the Writings of Poets and Novelists," *American Journal of Insanity*, 1 (July 1844), 9–46.

15. Although I suppose no one would deny that Melville's prose is sometimes overstrained and confused, there is room for basic disagreement about his style and narrative method. Richard H. Brodhead, for example, one of the ablest recent critics of Melville, while conceding that in the second half of *Pierre* the author grows "careless" and "introduces details that are utterly unprepared for and that scarcely fit together with what he has already told us," would apparently defend on philosophical grounds passages in *Moby-Dick* that strike me as unclear and contradictory. The book, he says, shows "a peculiar willingness to be in uncertainty, to embrace contradictions without resolving their antinomies . . ." (*Hawthorne, Melville and the Novel*, Chicago, 1976, pp. 184, 151). His guiding principle is as follows: "The only book that can faithfully represent a world that is various and mysterious is one that is susceptible to endless extension and modification. It must leave

room for each new aspect of things and each new motion of mind in a process of discovery that is, in the nature of things, still going on" (p. 15).

16. Paul Brodtkorb, Jr., *Ishmael's White World: A Phenomenological Reading of Moby-Dick*, New Haven, 1965, pp. 63-64.

17. Robert Zoellner, *The Salt-Sea Mastodon: A Reading of Moby-Dick*, Berkeley, 1973, pp. 96-101.

18. Note the difficulty one has in visualizing Ahab's being invited to leap into a chasm that is "in himself" yet "yawned beneath him."

19. Michel Foucault, in *Madness and Civilization: A History of Insanity in the Age of Reason* (trans. Richard Howard, New York, 1965), offers a rich and stimulating account of attitudes toward the insane in Europe from the Middle Ages to the nineteenth century.

20. William Charvat, "Melville and the Common Reader," in *Studies in Bibliography*, 12:49-50.

21. *Moby-Dick*, eds. Luther S. Mansfield and Howard P. Vincent, New York, 1952, pp. 576, 643.

22. Thomas Carlyle, *Sartor Resartus*, ed. Charles F. Harrold, New York, 1937, p. 260.

23. *Moby-Dick* (Mansfield-Vincent Edition), pp. 576, 643.

24. Or perhaps only one with two masks, as Staats in effect maintains in his contention that Ishmael represents the preliminary, "melancholic" or neurotic stage of a mental disease whose final stage is Ahab's monomania, a psychosis ("Melville—*Moby-Dick*," p. 123: "One may conclude that the two are one character or rather two phases of one imaginative conception, which originates in the experience of alienation. Ahab, the monomaniac, is the radicalization of the melancholic Ishmael, who from the beginning has shared with Ahab the vision of the White Whale that undertakes to destroy Ahab"). Again, this interpretation seems to me valuable for its linking of Ishmael and Ahab, but over-schematic, implying a greater degree of coherence in the book than it actually has.

25. Jean-Jacques Mayoux makes this point in one of the best critical essays about Melville ever published: "The *I* is the insignificant Ishmael. By the force of this *I*, the reader is led to identify Melville with Ishmael and the figure of Ahab is liberated; in relation to the author he has the independence and the authority of a vision. Looking at him more clearly, one sees readily that the suffering or damnation of Ahab is more important and comes nearer representing the choice of the author than the salvation of Ishmael. One notices that it is simply the schema of

Mardi in reverse and that Ishmael, grown wiser after he is rescued, will be ready to rejoin Babbalanja in some Serenia, while Ahab descends to rest in the bottomless pit beside the bones of Taji" ("Mythe et symbole chez Herman Melville," in *Vivants Piliers: le roman anglo-saxon et les symboles,* Paris, 1960, pp. 68–69; my translation).

26. Staats (p. 127) links this passage with the earlier passage (in Chap. 10, "A Bosom Friend," p. 53) that describes "the homoerotic, symbolic marriage of Ishmael with the savage who heals him of his melancholy," and cites also the homosexual coloring of Whitman's ideal of a radical democracy.

27. Charles Olson, *Call Me Ishmael,* p. 52; *Moby-Dick* (Mansfield-Vincent Edition), p. 643.

28. Merrell R. Davis, "Emerson's 'Reason' and the Scottish Philosophers," *New England Quarterly,* 17 (June 1944), 209–228.

29. Emerson, "Self-Reliance" (1840), in *Selections,* ed. Whicher, pp. 149–150.

30. Henry A. Murray, "In Nomine Diaboli," in *Moby-Dick Centennial Essays,* eds. Tyrus Hillway and Luther S. Mansfield, Dallas, 1953, pp. 13, 20. In his introduction to *Pierre,* Murray says that Melville "revealed the forces, antithetical to the contemporary cultural compound of puritanism, rationalism, and materialism, which were lurking, barbarized by repression, in the heart of Western Man, biding the moment for their eruption" *(Pierre,* New York, 1949, p. xxxi).

31. "Melville—*Moby-Dick,*" p. 113. Similar denunciations of modern society have been developed by Herbert Marcuse in *Eros and Civilization: A Philosophical Inquiry into Freud* (1955) and, with a stronger political emphasis, in *One-Dimensional Man: Studies in the Ideology of Advanced Industrial Society* (1964); and from a psychiatric standpoint by R. D. Laing in *The Divided Self: An Existential Study in Sanity and Madness* (1959) and *The Politics of Experience* (1967).—The dissemination of ideas of this kind in the United States (or at least in California) is indicated by an announcement in the *San Francisco Chronicle* (29 July 1968, p. 4, col. 5) that the Esalen Institute would sponsor two lectures on "The Science of Madness" as part of a one-month seminar series "The Value of Psychotic Experience." The faculty for the seminar included Dr. Stanislaf Grof (Prague), Julian Silverman (University of California, Santa Cruz), Allen Ginsberg, and Alan Watts.

32. Brodtkorb, *Ishmael's White World,* p. 127.

33. Michael P. Woolf, "The Madman as Hero in Contemporary American Fiction," *Journal of American Studies,* 10 (August 1976), 257–269.

Chapter 4   A Textbook of the Genteel Tradition:
             Henry Ward Beecher's *Norwood*

1. Marvin Felheim, "Two Views of the Stage; or, The Theory and Practice of Henry Ward Beecher," *New England Quarterly*, 25 (September 1952), 314-326.

2. Constance M. Rourke, *Trumpets of Jubilee: Henry Ward Beecher, Harriet Beecher Stowe, Lyman Beecher, Horace Greeley, P. T. Barnum*, New York, 1927, p. 187.

3. William G. McLoughlin, *The Meaning of Henry Ward Beecher: An Essay on the Shifting Values of Mid-Victorian America, 1840-1870*, New York, 1970, p. 252.

4. Robert Shaplen, *Free Love and Heavenly Sinners: The Story of the Great Henry Ward Beecher Scandal*, New York, 1954, pp. 18-20.— Dwight L. Moody and perhaps two or three other revivalists could draw larger crowds for a few days at a time (*Popular Culture and Industrialism 1865-1890*, ed. Henry N. Smith, New York, 1967, pp. 465-469).

5. To Edna Dean Proctor, Amesbury, Mass., 10 November 1860, in *Letters*, ed. John B. Pickard, 3 vols., Cambridge, Mass., 1975, 2:477.

6. Samuel L. Clemens, *The Love Letters of Mark Twain*, ed. Dixon Wecter, New York, 1947, pp. 53-55.

7. McLoughlin, *Meaning of Henry Ward Beecher*, pp. 10, 27, 57, 63; Shaplen, *Free Love and Heavenly Sinners*, pp. 18-20.

8. Unsigned review, "Beecher's *Norwood*," *Catholic World*, 10 (December 1869), 394.

9. Howells, review of *Famous Americans of Recent Times*, by James Parton, *Atlantic Monthly*, 19 (May 1867), 637.

10. To Celia Thaxter, Amesbury, Mass., 24 July 1870; to William Allinson, Amesbury, Mass., before 19 February 1870, *Letters*, 3:229; 3:215.

11. To Lydia Maria Child, Amesbury, Mass., 20 September 1874, *Letters*, 3:322.

12. Rourke, *Trumpets of Jubilee*, pp. 234-237.

13. Henry Ward Beecher, *Norwood; or, Village Life in New England*, New York, 1868, p. v.

14. Howells, review of *Norwood*, *Atlantic Monthly*, 21 (June 1868), 761.

15. *New Englander*, 27 (April 1868), 412, 411.

16. *Nation*, 6 (2 April 1868), 274-275.

17. Horace Bushnell, *Christian Nurture* (1847), especially the introduction by Luther A. Weigle in his edition (New Haven, Conn., 1947), pp. xxxi-xl.

18. *Nation*, 6 (2 April 1868), 275.

19. *Atlantic Monthly*, 21 (June 1868), 761. The reviewer for the *New Englander* also said that Beecher preaches too much, "though the preaching is much of it very good, and quite to our mind" (27 (April 1869), 412).

20. *Putnam's Magazine*, 1 (June 1868), 769–770.

21. *Harper's Monthly*, 36 (May 1868), 816.

22. Elizabeth Wetherell (pseud. of Susan B. Warner), *The Wide, Wide World* (1850), 2 vols., New York, 1852, 2:228.

23. Unsigned review, "Beecher's *Norwood*," *Catholic World*, 10 (December 1869), 394; unsigned article, "Beecherism and Its Tendencies," *Catholic World*, 12 (January 1871), 445, 448.

24. McLoughlin, *Meaning of Henry Ward Beecher*, pp. 255–256.

25. Robert Shaplen presents a detailed account of these events in *Free Love and Heavenly Sinners*.

26. "The Moral Background" (1920), in *The Genteel Tradition*, p. 85.

27. McLoughlin, *Meaning of Henry Ward Beecher*, pp. 64, 132.

28. Entry for 15 May 1873 in *Letters of Charles Eliot Norton*, eds. Sara Norton and M. A. DeWolfe Howe, 2 vols., Boston, 1913, 1:503–504.

29. Stephen E. Whicher, *Freedom and Fate: An Inner Life of Ralph Waldo Emerson* (1953; Philadelphia, 1957, pp. 109–111): "Although Emerson refused to conceive of life as tragedy, there is a sense in which his view of life can properly be called tragic, in so far as his recognition of the limits of mortal condition meant a defeat of his first romance of self-union and greatness."

30. Emerson, "Fate" (1860), in *Selections*, ed. Whicher, pp. 332, 333, 338.

31. Rose Wentworth summarizes the view of her father, who is Beecher's mouthpiece: ". . . one part of her truth Nature expresses to the senses, and another and far higher, through the senses she expresses to the soul" (*Norwood*, p. 205).

32. Emerson, "Nature" (1836), in *Selections*, pp. 35, 31.

33. "Realism" appears in this country in discussions of painting as early as the mid-1850s (Roger B. Stein, *John Ruskin and Aesthetic Thought in America, 1840–1900*, Cambridge, Mass., 1967, pp. 115–116), but was not applied to literature before the mid-1860s. The earliest use of the term in American literary criticism that I know of, which occurs in a review by Henry James in the *Nation* (1 (14 September 1865), 77), suggests the link between painting and fiction: "For an exhibition of the true realistic *chique* we would . . . compare that body of artists who are represented in France by MM. Flaubert and Gérôme to that

class of works which in our literature are represented by [Charlotte Yonge's] 'Daisy Chain' and 'The Wide, Wide World.'" (James evidently has in mind Aunt Fortune and Mr. Van Brunt.) But as late as 1869 the term "realism" had still not come into general use. Two different reviewers of Harriet Beecher Stowe's *Old Town Folks* referred to her portrayal of non-genteel characters as "the truth of Pre-Raphaelitism" (*Harper's Monthly*, 36 (August 1869), 454) and "a pre-Raphaelite picture" (*Galaxy*, 8 (July 1869), 142).

34. The moralistic tone of the terms "conscience" and "humility" anticipates Howells's attitude toward realism in literature.
35. Emerson, "Nature," in *Selections*, p. 26.
36. Erich Auerbach, *Mimesis: The Representation of Reality in Western Literature*, trans. Willard Trask, Princeton, N.J., 1953, pp. 40–46.
37. *Atlantic Monthly*, 21 (June 1868), 761.
38. *Nation*, 6 (2 April 1868), 275.
39. *Harper's Monthly*, 36 (May 1868), 816.
40. *Atlantic Monthly*, 21 (June 1868), 763.—Beecher himself loved fast horses. Whittier wrote to a friend concerning a visit to New York in 1859: "Henry Ward Beecher gave me a break-neck sort of drive over the city with his spirited team of horses . . ." (to Hannah J. Newhall, Amesbury, Mass., 17 May 1859, *Letters*, 2:407).
41. For example, New York *Tribune*, 10 (27 December 1850), p. 6.
42. *Literary World*, 28 (December 1850), 524.
43. Above, note 33.
44. *Atlantic Monthly*, 10 (July 1862), 126–127.
45. Auerbach, *Mimesis*, p. 491.
46. Daniel Aaron, *The Unwritten War: American Writers and the Civil War* (The Civil War Centennial Commission Series: The Impact of the Civil War, ed. Harold M. Hyman), New York, 1973.
47. Concord, 26 September 1864, in *The Correspondence of Emerson and Carlyle*, ed. Joseph Slater, New York, 1964, p. 542.
48. Review of *The Schönberg-Cotta Family*, *Nation* (14 September 1865), reprinted in *Notes and Reviews* by Henry James, ed. Pierre de Chaignon la Rose, Cambridge, Mass., 1921, pp. 81–82.
49. In an unpublished fragment called "The Start," written shortly after 1900, quoted in Henry Nash Smith, *Mark Twain's Fable of Progress*, New Brunswick, N.J., 1964, p. 93.
50. Ursula Brumm, *American Thought and Religious Typology*, New Brunswick, N.J., 1970, p. 217.
51. Ibid., p. 220.

Chapter 5 William Dean Howells: The Theology of Realism

1. Kenneth Lynn, *William Dean Howells: An American Life*, New York, 1971, pp. 66–127.
2. Nathaniel Hawthorne, *The Marble Faun*, p. 3.
3. William Dean Howells, *Venetian Life*, New York, 1867, pp. 103 ff.
4. Relevant passages in *The Innocents Abroad* are Mark Twain's description of the visit to Pompeii, in which the whistle of a railway locomotive seems a shocking intrusion of modern civilization into the dreamy world of the past (2 vols., New York, 1911, Vol. 2, Chap. 4), and the monologue of the peasant from the Campagna who has returned from viewing the wonders of civilization in the United States (1, Chap. 26). The popular contrast Progress-Efficiency-Work-Rationality-Realism-as-truth-telling versus Backwardness-Indolence-Dreams-Sentiment-Romance-as-distorting-fantasy would be developed more fully in *Life on the Mississippi* (1883) when the narrator travels from the deep South of Louisiana, where the baneful influence of Walter Scott still fosters a decadent cult of chivalry, to the upper Middle West, where the brisk, prosperous cities along the River represent the cutting edge of progress and modernity.
5. In a review of Howells's *A Foregone Conclusion* (*North American Review*, 120 (January 1875), 213), Henry James expresses regret that at the very end of the story the author moves his characters from the "ideal" setting of Venice to the "real, to the vulgar" setting of the United States.
6. *North American Review*, 103 (October 1866), 611.
7. Quoted in Charles E. Stowe, *Life of Harriet Beecher Stowe*, Boston, 1890, p. 334. —Lowell's assertion that the ideal lies in "the simply natural" closely resembles Howells's doctrine of realism as it would take shape some ten years later. Further: Lowell assumes a disjunction between "ideal interest" and "character." "Character" is identified with "nature," and both have approximately the meaning that Howells would later assign to the term "reality."
8. William Dean Howells, reviews of Arne: *The Happy Boy; The Fisher Maiden*, in *Atlantic* (April 1870), reprinted in *Criticism and Fiction and Other Essays*, eds. Clara M. Kirk and Rudolf Kirk, New York, 1959, p. 106.
9. Anon., "Mr. Howells His Own Critic," *Literary News* (New York), N.S. 18 (October 1897), p. 313. Howells added that *Indian Summer* (1886) was "The one I *like* best"; and he omitted all mention of the two of his books

now most often read—namely, *The Rise of Silas Lapham* (1885) and *A Hazard of New Fortunes* (1890).

10. Howells, *Criticism and Fiction*, pp. 49, 47. Howells often restated this principle—for example, in a comment on Mark Twain in *Harper's Monthly* in 1887 (74 (May), 987): "Let fiction cease to lie about life; let it portray men and women as they are, actuated by the motives and the passions in the measure we all know; let it leave off painting dolls and working them by springs and wires. . . ."

11. Henry F. May, *The End of American Innocence*, New York, 1959, p. 14.

12. Lynn, *Howells*, pp. 34-35, 41.

13. In an unusually interesting article called "Agnosticism in American Fiction," published in 1884 in the theologically oriented *Princeton Review* (Series 4: Vol. 13 (January), 1-15). Julian Hawthorne placed both Howells and Henry James among a new generation of novelists exhibiting a tendency toward agnosticism, and asserted that "Agnosticism . . . reaches forward into nihilism on one side and extends back into liberal Christianity on the other . . ." (p. 3). —May observes that Howells's "smiling realism" was "the exact literary version of practical idealism" (*End of American Innocence*, p. 16).

14. William Dean Howells, *A Modern Instance* (Riverside Edition), ed. William M. Gibson, Boston, 1957, p. 10. Subsequent citations to this edition are included within parentheses in the text.—For later reference, let me note here how Henry James has Gilbert Osmond make ironic use of a similar ploy with Isabel Archer: "It has made me better, loving you . . . it has made me wiser and easier and—I won't pretend to deny—brighter and nicer and even stronger" (*The Portrait of a Lady*, Modern Library Edition, New York, n.d., 2 vols., 2:81).

15. Eugen Lerch, "Ursprung und Bedeutung der sogenannten 'Erlebten Rede' ('Rede als Tatsache')," in *Hauptprobleme der Französischen Sprache* (Band I), Braunschweig, 1930, pp. 91-138. Stephen Ullmann discusses Flaubert's use of represented discourse at greater length in *Style in the French Novel*, Cambridge (Eng.), 1957, Chap. 2.

16. It is also significant that the story was suggested to Howells by a performance of a translation of Franz Grillparzer's *Medea* (Gerard M. Sweeney, "The *Medea* Howells Saw," *American Literature*, 42 (March 1970), 83-89). The text was published in German with an anonymous English translation as *Medea, A Tragedy in Four Acts by Grillparzer, as performed by Mlle. Fanny Janauschek and her company of German artists, at the Academy of Music under the Direction of Max Maretzek*, New York, n.d.—Whereas Bartley often calls to mind a picaresque rogue, Marcia is modeled to some extent on a heroine of tragedy. Her rustic

igenuousness might seem from the perspective of Boston a kind of barbarism, and she has traces of passion on a tragic scale, as well as an aristocratic scorn of calculation.

17. Editor's note, *A Modern Instance*, p. 24. "Slang" had connotations of cynicism and "irreverence" that it has since lost.
18. See above, p. 70.
19. Auerbach, *Mimesis*, p. 491.
20. *Harper's Monthly*, July 1890, p. 318, quoted by William Alexander, "Howells, Eliot and the Humanized Reader," in *The Interpretation of Narrative: Theory and Practice* (Harvard English Studies, 1), ed. Morton W. Bloomfield, Cambridge, Mass., 1970, p. 153.
21. Howells, *My Literary Passions*, New York, 1895, pp. 169–70.
22. In an often anthologized passage from *Criticism and Fiction*, written a decade later, Howells would include "self-devoted" in a list of epithets belonging to the romantic mode of fiction he proposed to reject (p. 13). Other epithets on his list are "ideal," "heroic," "impassioned," and "adventureful."
23. Angels and devils appeared as allegorical figures in the Swedenborgian faith of Howells's father (Howard M. Munford, "The Genesis and Early Development of the Basic Attitudes of William Dean Howells," unpublished dissertation, Harvard, 1950, p. 37). When Bartley quarrels with his employee, the narrative voice says that "The demons, whatever they were, of anger, remorse, pride, shame, were at work in Bartley's heart too . . ." (p. 56). In a volume of reminiscences Howells wrote of his childhood: "The children were taught when they teased one another that there was nothing the fiends so much delighted in as teasing. When they were angry and revengeful, they were told that now they were calling evil spirits about them, and that the good angels could not come near them if they wished while they were in that state" (*A Boy's Town*, New York, 1890, p. 12).
24. Quoted by Gibson in his introduction to *A Modern Instance*, p. vi.
25. Quoted from an unpublished letter to Howells (dated 19 August [1882], MS at Harvard) by Olov W. Fryckstedt, *In Quest of America: A Study of Howells' Early Development as a Novelist*, Cambridge, Mass., 1958, p. 242.
26. Hartford, 24 July 1882, in *Mark Twain-Howells Letters*, 1:407–408, 412.
27. Lynn, *William Dean Howells*, p. 256.
28. In *Criticism and Fiction* Howells declared: ". . . it is the notion that a novel may be false in its portrayal of causes and effects that makes literary art contemptible even to those whom it amuses, that forbids them to regard the novelist as a serious or right-minded person" (p. 49).

29. Rodney M. Baine, *Daniel Defoe and the Supernatural*, Athens, Ga., 1968, esp. Chaps. 1 and 2.

30. Quoted by Jessica Mitford, *Kind and Usual Punishment: The Prison Business*, New York, 1973, p. 46, from Edwin H. Sutherland, *Principles of Criminology*, Chicago, 1934, p. 43. Further, "The Supreme Court of North Carolina declared in 1862: 'To know the right and still the wrong pursue proceeds from a perverse will brought about by the seductions of the Evil One.'"

31. Cooper, *The Pioneers* (Signet Edition), New York, 1964, p. 360.

32. Satan whispers in Dimmesdale's ear when Dimmesdale returns from his encounter with Hester Prynne in the forest (Hawthorne, *Scarlet Letter*, pp. 220–222), and angels weep while fiends rejoice as Dimmesdale stands on the scaffold at night (p. 148). It is rumored that the fire in Chillingworth's laboratory "had been brought from the lower regions, and was fed with infernal fuel" (p. 127).

33. *Lippincott's*, 2 (September 1868), 248.

34. *Atlantic*, 33 (March 1874), 370–371.

35. *Criticism and Fiction*, p. 43.

36. *William Dean Howells*, p. 254.

37. Howells to James R. Osgood, 18 February 1881, quoted in Gibson, "Introduction," *A Modern Instance*, p. vii.

38. Ibid., pp. viii–ix.

39. Erik H. Erikson, "Autobiographic Notes on the Identity Crisis," *Daedalus*, 99 (Fall 1970), 730–759; quotation p. 753.

40. *William Dean Howells*, p. 254. Washington Gladden, a nationally known and respected liberal clergyman, contributed to the *Century* while *A Modern Instance* was appearing serially in that magazine (but before the appearance of the last section, with the debates between Atherton and Halleck) an article entitled "The Increase of Divorce" that is much more moderate in tone than Atherton's pronouncements. Gladden quotes Sir Henry Maine to the effect that the changes in laws governing divorce in the United States during the past thirty years were part of a general "movement from status to contract" observable also in Europe. The clergyman deplores the increase in divorces and observes that on the whole the recent changes in laws on the topic have produced "more domestic unhappiness than they have prevented." But he himself advocates allowing divorce for desertion "after a long term of years" (23 (January 1882), 411–420). It is worth noting that in *Miss Ravenel's Conversion from Secession to Loyalty*, by John W. De Forest (published in 1867)—a book that Howells praised highly—the protagonist Edward Colbourne is in love with Lily Ravenel before and

during her marriage to the dashing but immoral John Carter, yet the author marries him to Lily after Carter's death in battle without seeming to be aware of any moral problems.

41. *Century*, 25 (January 1883), 463-464.

42. *Nation*, 36 (January 1883), 41.

43. Although Howells keeps the sexual component in Hubbard's and Halleck's attitudes toward Marcia deeply buried, an additional contrast between them appears in Bartley's reverie (in Passage A, above) and the narrative voice's comment on Halleck's reaction to his first sight of Marcia's infant daughter: "There is something in a young man's ideal of women, at once passionate and ascetic, so fine that any words are too gross for it" (p. 196).

44. Ben Halleck reflects briefly, in Atherton's dining room, that Hubbard "might never have gone wrong if he had had all that luxury," but the narrative voice calls this "a perverse sympathy with Bartley," and the suggestion is not developed (p. 325). An anonymous editorial in *Century* (24 (October 1882), 941)—the issue in which the last installment of the novel appears—asserts that "Atherton, the clear-headed, clean-hearted lawyer" utters the central "moral meaning" of *A Modern Instance* in the speech of which the relevant part is quoted as Passage G above.

45. Horace Bushnell, *Christian Nurture*, New York, 1916, p. 210. The Puritan background of Bushnell's doctrine is richly documented in Norman Pettit, *The Heart Prepared: Grace and Conversion in Puritan Spiritual Life* (Yale Publications in American Studies, 11), New Haven, 1966.

46. This set of ideas had been in circulation since the eighteenth century, and was especially powerful in the United States (J. G. A. Pocock, "Civic Humanism and Its Role in Anglo-American Thought," in *Politics, Language and Time: Essays on Political Thought and History*, New York, 1971, pp. 80-103; Charles A. Beard and Mary R. Beard, *The American Spirit*, New York, 1942, passim).

47. Howells, *A Hazard of New Fortunes*, eds. Don L. Cook and David J. Nordloh, Bloomington, Ind., 1976, p. 452.

Chapter 6   Guilt and Innocence in Mark Twain's Later Fiction

1. Kenneth S. Lynn, *William Dean Howells*, pp. 128-325.

2. Frank L. Mott, *A History of American Magazines* (5 vols., Cambridge, Mass., 1938-68), presents in vols. 2 and 3 brief individual histories of these magazines. Clarence Gohdes's chapter on "The Age of the

Monthly Magazines" (in Arthur H. Quinn, ed., *The Literature of the American People*, New York, 1951, pp. 569-597) is the most helpful over-all view.

3. The *Ladies' Home Journal*, for example, founded in 1883, reached a circulation of 700,000 by the early 1890s (ibid., p. 595).

4. Ibid., pp. 590-591. The circulation of *Harper's Weekly* is estimated from data for other years in Mott, *History of American Magazines*, 2:480-482.

5. Hjalmar H. Boyesen, "Why We Have No Great Novelists," *Forum*, 2 (February 1887), 615-622.

6. The novelist who would have his books serialized in the monthly magazines, declared Boyesen, "shuns large questions and problems because his audience is chiefly interested in small questions and problems. He avoids everything which requires thought, because, rightly or wrongly, thought is not supposed to be the ladies' *forte*. Their education has not trained them for independent reflection. . . . the novelist who aspires for their favor becomes, also, conservative, and refrains from discussing what, according to the boarding-school standard, is unsafe or improper" (ibid., 2:617).—Marianna Debouzy presents an especially vigorous brief description of the situation of the artist in the 1880s and 1890s in *La Genèse de l'esprit de révolte dans le roman américain, 1875-1915*, Paris, 1968, pp. 113-121.

7. Clemens to William Dean Howells, Hartford, 8 December 1874, in *Mark Twain-Howells Letters*, 1:49.

8. Arthur Scott, "*The Century Magazine* and *Huckleberry Finn*, 1884-1885," *American Literature*, 27 (November 1955), 356-362.

9. Henry N. Smith, "'That Hideous Mistake of Poor Clemens's,'" *Harvard Library Bulletin*, 9 (Spring 1955), 145-180.

10. Hamlin Hill, *Mark Twain and Elisha Bliss*, Columbia, Missouri, 1964, pp. 2-3.

11. Ibid., Chap. 1 ("The Only Way To Sell a Book"), esp. pp. 12-13.

12. Ibid., p. 19. More than twenty years later, Mark Twain wrote to his friend and business adviser Henry H. Rogers: "I think my reason for wanting my travel-book [*Following the Equator*] in Bliss's hands is sound. Harper publishes very high-class books and they go to people who are accustomed to read. That class are surfeited with travel-books. But there is a vast class that isn't—the factory hands and the farmers. *They* never go to a bookstore; they have to be hunted down by the canvasser. When a subscription book of mine sells 60,000, I always think I know whither 50,000 of them went. They went to people who don't visit bookstores.

"I planned this book, from the beginning, for the *subscription* market. I am writing it according to that plan" (London, November (1896), in *Mark Twain's Letters to His Publishers*, ed. Hamlin Hill, Berkeley, 1967, p. 7). Elisha Bliss's son, Francis E. Bliss, had succeeded his father as president of the firm.—In 1890, in a letter addressed to Andrew Lang concerning Lang's disapproval of *A Connecticut Yankee in King Arthur's Court*, Clemens offered a different statement of his intention as a writer, as unconvincing in its way as is the hard-boiled commercialism of the letter to Rogers: "I have never tried in even one single instance, to help cultivate the cultivated classes. . . . I never had any ambition in that direction, but always hunted for bigger game— the masses. I have seldom deliberately tried to instruct them, but have done my best to entertain them. To simply amuse them would have satis-fied my dearest ambition at any time; for they could get instruction else-where, and I had two chances to help to the teacher's one; for amuse-ment is a good preparation for study and a good healer of fatigue after it. My audience is dumb, it has no voice in print, and so I cannot know whether I have won its approbation or only got its censure" (*Mark Twain's Letters*, ed. Albert B. Paine, 2 vols. (paged continuously), New York, 1917, p. 528).

13. Hill, *Mark Twain and Elisha Bliss*, p. 11.
14. Ernest Hemingway, *Green Hills of Africa*, New York, 1935, p. 22.
15. Santayana, *The Genteel Tradition*, pp. 51–52.
16. *Mark Twain: The Development of a Writer*, Cambridge, Mass., 1962.
17. Santayana, *The Genteel Tradition*, pp. 51–52.
18. Ibid., p. 52. Santayana is apparently quoting, from memory, the open-ing of Mark Twain's *Burlesque Autobiography*, first published in 1871 and never reprinted in this country, although it was later included in several English editions of *Sketches New and Old*. The passage reads: "My parents were neither very poor nor conspicuously honest." San-tayana's revision of the passage seems to me a decided improvement.
19. *Adventures of Huckleberry Finn (Tom Sawyer's Comrade)* (1885), fac-simile text in *The Art of Huckleberry Finn*, eds. Hamlin Hill and Walter Blair, San Francisco, 1962, p. 270. Later citations within paren-theses in the text.
20. Notebook #28a (I), TS, p. 35 (1895), Mark Twain Papers.
21. "The Form of Freedom in *Adventures of Huckleberry Finn*," *Southern Review*, New Series, 6 (Autumn 1970), 960.
22. Walter Blair, *Mark Twain & Huck Finn*, Berkeley, 1960, p. 353.
23. Hartford, 17 November, in *Mark Twain-Howells Letters*, 1:279.
24. Ibid., 1:281 n. 3.

25. Although Mark Twain understandably fills out his image of Southern culture of the 1840s with items appropriate to the setting specified on the title page of *Adventures of Huckleberry Finn* ("The Mississippi Valley forty to fifty years ago"), many of the items portrayed had in the past been common to all parts of the country. For example, since Emmeline Grangerford's works of art are so conspicuous in the masterly depiction of the Grangerford household, it is perhaps worthy of mention that the household of the Reverend Mr. Grant, Episcopal minister of the village of Templeton, New York, in the 1790s as Cooper describes it in *The Pioneers*—a man for whom the author shows great respect—was ornamented by "specimens of needlework and drawing, the former executed with great neatness, though of somewhat equivocal merit in their designs, while the latter were strikingly deficient in both." One of the examples of needlework "represented a tomb, with a youthful female weeping over it, exhibiting a church with arched windows in the background. On the tomb were the names, with the dates of the births and deaths, of several individuals, all of whom bore the name of Grant" (Chap. 12, Signet Edition, p. 134).

26. Horst H. Kruse, *Mark Twains* Life on the Mississippi: *Eine entstehungs- und quellengeschichtliche Untersuchung zu Mark Twains* Standard Work (*Kieler Beiträge zur Anglistik und Amerikanistik*, Band 8), Neumünster, 1970.

27. Henry N. Smith, *Mark Twain's Fable of Progress*, pp. 48-49.

28. Since so much emphasis has been placed on the notion that at the end of *Adventures of Huckleberry Finn* Huck announces his permanent flight into a vague Western Territory, it is perhaps appropriate to note here that Mark Twain himself imagined Huck, Jim, and Tom as returning to St. Petersburg after a brief excursion out the Oregon Trail. *Tom Sawyer Abroad* (composed 1893/1894) and "Tom Sawyer, Detective" (composed 1896) make this assumption (Walter Blair, "Introduction," in *Mark Twain's Hannibal, Huck, & Tom*, Berkeley, 1969, pp. 9-20). But these attempts at a sequel to *Adventures of Huckleberry Finn* demonstrate that the Matter of Hannibal had lost its power to stir the writer's imagination deeply.

29. Robert Penn Warren, in *Democracy and Poetry* (The 1974 Jefferson Lecture in the Humanities, Cambridge, Mass., 1975), writes: "As Huck gradually develops a new 'consciousness' to replace the old 'conscience,' the reader's expectation rises that Huck may find for himself—and for the reader—a way to redeem life ashore, to create a life in which the 'real' of the shore and the 'ideal' of the river may meet, or at least enter into some fruitful relation. But nothing of the sort occurs" (p. 18).

30. *A Connecticut Yankee in King Arthur's Court* (facsimile edition), ed. Hamlin Hill, San Francisco, 1963, p. 20.

31. Chadwick Hansen calls attention to some surprising parallels between Hank Morgan's policies and program and those of twentieth-century dictators such as Mussolini and Hitler ("The Once and Future Boss: Mark Twain's Yankee," *Nineteenth-Century Fiction,* 28 (June 1973), 62-73). John F. Kasson develops more fully the insensitivity that Hank displays toward the victims of his modern weapons: "He delights in his power to kill efficiently and distantly through his technology and indulges in a feeble military wit, what might be called the pornography of destructive power, which conflates extraordinary and deadly actions with mundane and innocent ones" (*Civilizing the Machine: Technology and Republican Values in America 1776-1900,* New York, 1976, p. 213). And there is an unpleasant relish of violence in a passage in Mark Twain's manuscript, fortunately omitted from the published text, in which the Yankee engages in calculations that enable him to determine that "we had killed 1,069,362 pounds." This leads him to comment: "We were very well pleased with the result. An entire fifth part of the enemy had been disposed of in this engagement. The gross weight of the remaining four-fifths was estimated at 4,277,448 pounds" (Emendations List in Bernard L. Stein's edition of *A Connecticut Yankee,* in press with University of California Press, Berkeley).

32. Smith, *Mark Twain's Fable of Progress,* pp. 76-82. Reviewers of *A Connecticut Yankee* were perhaps influenced by the fact that the author of that book was a celebrated humorist. Particularly relevant is the "Raft Passage" that he lifted from the manuscript of *Adventures of Huckleberry Finn* and inserted in *Life on the Mississippi.* In that episode a raftsman who calls himself the Child of Calamity shouts (rather improbably): "I'm the man with a petrified heart and biler-iron bowels! The massacre of isolated communities is the pastime of my idle moments, the destruction of nationalities the serious business of my life! The boundless vastness of the great American desert is my enclosed property, and I bury my dead on my own premises!" (*Life on the Mississippi,* ed. J. C. Levenson, Minneapolis, 1967, p. 19).

33. William M. Gibson, "Introduction," in *Mark Twain's Mysterious Stranger Manuscripts,* Berkeley, 1969, p. 17.—The discussion which follows draws on Paul Baender's unpublished dissertation, "Mark Twain's Transcendent Figure," University of California (Berkeley), 1956.

34. Ibid., pp. 4-5; John S. Tuckey, *Mark Twain and Little Satan: The Writing of* The Mysterious Stranger, (Purdue University Studies), West Lafayette, Ind., 1963.

35. *Mysterious Stranger Manuscripts*, pp. 15-16.
36. Ibid., p. 16.
37. Ibid., p. 73.
38. Ibid., p. 405.

Chapter 7   Henry James I: Sows' Ears and Silk Purses

1. William Veeder, *Henry James—The Lessons of the Master: Popular Fiction and Personal Style in the Nineteenth Century*, Chicago, 1975, p. 18.
2. Many of James's early reviews of fiction are collected in *Notes and Reviews*, ed. Pierre de Chaignon la Rose, Cambridge, Mass., 1921.
3. Veeder, *Lessons of the Master*, p. 8.
4. Henry James, *Washington Square*, ed. Clifton Fadiman, New York, 1950, p. 13.
5. Veeder, *Lessons of the Master*, p. 77.
6. Henry James, *The Princess Casamassima* (New York Edition), 2 vols., New York, 1908: literature (1:315, 2:343, 276); women's dress (1:166); painting (2:123); interior decoration (1:159). Hyacinth's "mixed blood" is mentioned at 2:264.
7. James, *Princess*, 1:4-5, 2:343.
8. The hero is hurled over a precipice but reappears later unharmed (1:213). The last act contains a scene with pistol shots and shrieks (1:225).
9. Henry James, *In the Cage* (1898), London, 1919, p. 9.
10. *The Art of the Novel: Critical Prefaces by Henry James*, ed. Richard P. Blackmur, New York, 1962, pp. 57, 257-258. The repeated references to pirates suggest that James may have been recalling tales he had read in his boyhood, which coincided in time with Tom Sawyer's.
11. James to Howells, Paris, 24 October 1876, in *Henry James Letters*, ed. Leon Edel, 2 vols., Cambridge, Mass., 1974-1975, 2:70.
12. London, 30 March 1877, *Letters*, ed. Edel, 2:104-105.
13. Milford, Godalming, Surrey, 30 October 1878, *Letters*, ed. Edel, 2:189.
14. Leon Edel, *Henry James: The Treacherous Years, 1895-1901*, Philadelphia, 1969, pp. 155-156. Edel points out that James sold the article to the New York *Herald* and the Boston *Herald*, both of which published it "under large headlines, as if indeed James had written a shocker." The essay was published also in the (English) *New Review*. The implication must be that the literary monthlies maintained a particularly rigid code of propriety. Howells wrote in 1891: "Between

the editor of a reputable English or American magazine and the families which receive it there is a tacit agreement that he will print nothing which a father may not read to his daughter, or safely leave her to read herself" (*Criticism and Fiction and Other Essays*, p. 75).

15. *The Complete Tales of Henry James*, ed. Leon Edel, 12 vols., Philadelphia, 1962-1964, 9:424.

16. James to Howells, Paris, 21 February 1884, in *The Letters of Henry James*, ed. Percy Lubbock, 2 vols., New York, 1920, 1:104. At about the same time James has Mark Ambient, the author of *Beltraffio*, denounce his philistine wife's taste in comparably violent language: ". . . her conception of a novel . . . is a thing so false that it makes me blush. It is a thing so hollow, so dishonest, so lying, in which life is so blinked and blinded, so dodged and disfigured, that it makes my ears burn" (*Complete Tales*, 5:336). Many years later James expressed a comparable emotion, in slightly more temperate language, in one of his "American Letters" to the London journal *Literature*, concerning the large sales of Paul Leicester Ford's *The Honorable Peter Stirling* (7 May 1898, reprinted in *Henry James. The American Essays*, ed. Leon Edel, New York, 1956, pp. 223-224).

17. London, 2 January 1888, in *Letters*, ed. Lubbock, 1:135. James had told his brother William that *The Bostonians* "is, I think, the best fiction I have written. . ." (London, 14 February 1885, 1:117).

18. London, 12 January 1891, in *Letters*, ed. Lubbock, 1:176.

19. London, 5 August 1893, in *Letters*, ed. Lubbock, 1:206-207.

20. Edel, *Henry James: The Treacherous Years*, pp. 72-80.

21. James, *Complete Tales*, 8:436.

22. T. S. Eliot, "On Henry James," reprinted in *The Question of Henry James. A Collection of Critical Essays*, ed. Frederick W. Dupee, New York, 1945, p. 110: "James's critical genius comes out most tellingly in his mastery over, his baffling escape from, Ideas; a mastery and an escape which are perhaps the last test of a superior intelligence. He had a mind so fine that no idea could violate it." (Originally published in the *Little Review*, August 1918.)

23. Entry for 26 January 1895, in *The Notebooks of Henry James*, eds. F. O. Matthiessen and Kenneth B. Murdock, New York, 1947, p. 180.

24. Entry for 4 June 1895, in *Notebooks*, pp. 202, 204. Thirty years earlier James had observed that Miss Braddon had "created the sensation novel" (with *Lady Audley's Secret*, 1862) (*Notes and Reviews*, p. 109). Edel says that Mrs. Highmore satirizes Mrs. Humphrey Ward *(Treacherous Years*, p. 293).

25. Ilse Dusoir Lind, "The Inadequate Vulgarity of Henry James," *PMLA*,

66 (December 1951), 886–910. Even before Reid suggested a change of
approach, James had written to his father that "the vulgarity and
repulsiveness of the Tribune, whenever I see it, strikes me so violently
that I feel tempted to stop my letter" (*PMLA*, 66:896). And in recalling
his early relation with the *Nation* after the lapse of forty years James
said that "though I suppose I should have liked regularly to correspond
from London, nothing came of that but three or four pious efforts
which broke down under the appearance that people liked most to hear
of what I could least, of what in fact nothing would have induced me
to, write about. What I could write about they seemed, on the other
hand, to view askance . . ." ("The Founding of the Nation," 101
(8 July 1915), 44–45). Given the high intellectual level of the *Nation*,
this makes one wonder whether James is not rationalizing somewhat
the fact that he was temperamentally unsuited to ordinary journalism.
In a later story ("The Papers," 1903), a sympathetic young newspaper-
man says to an equally sympathetic young newspaperwoman: ". . . we
know we've got to live, and how we do it. But at least, like this, alone
together, we take our intellectual revenge, we escape the indignity of
being fools dealing with fools. I don't say we shouldn't enjoy it more
if we *were*. But it can't be helped; we haven't the gift—the gift, I mean,
of not seeing. We do the worst we can for the money" (*Complete Tales*,
12:21).

26. James, *Complete Tales*, 9:187. It is worth noting that James transfers
his experience with a New York paper to Britain: he has ceased distin-
guishing significantly in his own mind between the two reading pub-
lics. In 1888 James had written to his brother William: "I can't look at
the English-American world, or feel about them, any more, save as a big
Anglo-Saxon total, destined to such an amount of melting together that
an insistence on their differences becomes more and more idle and
pedantic . . ." (Geneva, 29 October 1888, in *Letters*, ed. Lubbock,
1:141).

27. See above, pp. 10–11. It will be recalled that Melville believed Hawthorne had managed something like this in "Young Goodman Brown"
and other sketches. James, interestingly enough, implies that in
Howells's *A Chance Acquaintance* he "managed at once to give his
book a loose enough texture to let the more simply-judging kind fancy
they were looking at a vivid fragment of social history itself, and yet
to infuse it with a lurking artfulness which should endear it to the
initiated" (*Literary Reviews and Essays*, ed. Albert Mordell, New York,
1957, p. 203; first published in the *North American Review*, January
1875).

Chapter 8   Henry James II: The Problem of an Audience

1. Edel, *Henry James: The Master, 1901–1916*, Philadelphia, 1972, p. 436.
2. Ibid., pp. 476–477.
3. Allen Walker Read, "The Membership in Proposed American Academies," *American Literature*, 7 (May 1935), 155, 157, 159.
4. London, 25 August 1915, in *Letters*, ed. Lubbock, 2:497.
5. See Chapter 4, note 33, above.
6. Among many essays published in nineteenth-century journals, I have found the following particularly useful: James Herbert Morse, "Henry James, Jr., and the Modern Novel," *Critic*, 2 (14 January 1882), 1. Charles Dudley Warner, "Modern Fiction," *Atlantic*, 51 (April 1883), 464–474. Julian Hawthorne, "Agnosticism in American Fiction," *Princeton Review*, 4th ser., 13 (January 1884), 1–15. Arlo Bates, "Realism and the Art of Fiction," *Scribner's*, 2 (August 1887), 241–252.— I wish to acknowledge with thanks the indispensable assistance of Robert H. Hirst in gathering materials for the survey of reviews of Henry James's books that occupies most of the present chapter.
7. William Dean Howells, "Henry James, Jr.," *Century*, 25 (November 1882), 25–29.
8. [James Russell Lowell], "James's Tales and Sketches," *Nation*, 20 (24 June 1875), 425–427.
9. [Anon. rev. of *Roderick Hudson*], *Appleton's Journal*, 14 (18 December 1875), 793.
10. Henry James wrote to Mrs. Sarah Butler Wister (mother of Owen Wister) in 1874 concerning *Roderick Hudson:* "The fault of the story [i.e., according to reviewers], I am pretty sure, will be in its being too analytical and psychological, and not sufficiently dramatic and eventful" (Baden-Baden, 29 July 1874, in *Letters*, ed. Edel, 1:460).
11. *North American Review*, 122 (April 1876), 420–424.
12. *North American Review*, 128 (January 1879), 101. William C. Brownell intended to express the same opinion when he wrote that "Confidence" is "a study rather than a story" (*Nation*, 30 (25 March 1880), 239).
13. *Californian*, 3 (April 1881), 376–377.
14. *Critic*, 2 (14 January 1882), 1.
15. *Literary World*, 12 (17 December 1881), 473–474. An anonymous reviewer of George Eliot's *Daniel Deronda* in the *Nation* (23 (19 October 1876), 245) had complained that "There is something absolutely painful in the kind of vivisection to which his [Deronda's] physical and moral qualities are subjected."

16. Brownell recognized that James's distinction "consists in his attempt to dispense with all the ordinary machinery of the novelist except the study of subtle shades of character" ("James's *Portrait of a Lady*," *Nation*, 34 (2 February 1882), 102–103).

17. "Agnosticism in American Fiction," *Princeton Review*, 4th ser., 13 (January 1884), 1–15.

18. Oscar Cargill, *The Novels of Henry James*, New York, 1961, pp. 131–132.

19. James to Edmund Gosse, London, 25 August 1915, in *Letters*, ed. Lubbock, 2:498.

20. *Dial*, 7 (May 1886), 15. It is baffling to notice that later in the same year, in a review of *The Princess Casamassima*, Payne referred to "the admirable analytic quality of the work of Mr. James" (*Dial*, 7 (December 1886), 189).

21. *Nation*, 44 (10 February 1887), 123–124.

22. *Independent*, 38 (22 April 1886), 495.

23. *Independent*, 38 (23 December 1886), 1665.

24. *Lippincott's*, 39 (February 1887), 359.

25. *Critic*, 10 (29 January 1887), 51–52.

26. E.g., William Morton Payne, reviewing *The Tragic Muse* in the *Dial* (11 (August 1890), 93). James's popularity had reached a low point in Britain toward the end of the 1880s (Donald M. Murray, "Henry James and the English Reviewers, 1882–1890," *American Literature*, 24 (March 1952), 1–20, esp. 18–20).

27. *Literary World*, 24 (8 April 1893), 113.

28. *Dial*, 14 (1 June 1893), 341.

29. *Critic*, 23 (14 October 1893), 236.

30. *Nation*, 57 (30 November 1893), 416–417.

31. *Critic*, 23 (21 October 1893), 253.

32. *Dial*, 15 (December 1893), 344.

33. *Outlook*, 55 (27 February 1897), 610.

34. *Public Opinion*, 22 (11 March 1897), 312.

35. New York *Times, Saturday Review of Books and Art*, 2 (20 February 1897), 1.

36. "Mr. James's 'Adorable Subtleties,'" *Book Buyer*, 14 (April 1897), 303–305.

37. *American*, 3 (31 December 1881), 186–187, quoted in Linda J. Taylor, "*The Portrait of a Lady* and the Anglo-American Press: An Annotated Checklist, 1880–1886," *Resources for American Literary Study*, 5 (Autumn 1975), 181.

38. *Literary World*, 16 (21 March 1885), 102.

39. *Independent*, 49 (16 December 1897), 1660.

40. *Bookman*, 6 (February 1898), 562.

41. *Public Opinion*, 23 (30 December 1897), 855.

42. *Outlook*, 57 (13 November 1897), 670.

43. *Critic*, 32 (8 January 1898), 21.

44. *Nation*, 66 (17 February 1898), 135. The anonymous reviewer was apparently not Annie R. M. Logan.

45. *Sewanee Review*, 8 (January 1900), 112-113.

46. *Outlook*, 66 (13 October 1900), 423.

47. The eventual acceptance of James could be attributed (as it is in an anonymous review of *The Outcry* in the *North American Review* in 1912) to the recognition that his work is "as truly scientific as it is finely literary. . . ." The reviewer notes that "It has been said that Mr. James relies too much upon the head to the exclusion of the heart, that there is too great intellectual astringency in his work as compared with its emotional quality. . . ." But "In this last novel . . . the rights of head and heart are delightfully balanced" (195 (January 1912), 141-143).

48. *Harper's Monthly*, 102 (May 1901), 974.

49. New York *Times, Saturday Review of Books and Art*, 7 (4 October 1902), 658. Similar observations were being made in England. In reviewing *The Wings of the Dove*, the (London) *Times Literary Supplement*, although declaring that James "is to be congratulated," acknowledged that "it is not an easy book to read. . . ." Of James's novels generally, the *TLS* declared that "the average reader misses in them . . . a sense of good-fellowship, and a common attitude towards life. . . . There is something of the classic in his [James's] sense of aloofness, his detachment from his reader; and the pampered modern reader is apt to call the attitude inhuman" (1 (September 1902), 263).

50. *Independent*, 54 (13 November 1902), 2711-2712.

51. "Henry James's Short Stories," *Lamp [Book Buyer]*, 26 (April 1903), 232.

52. *Current Literature*, 34 (May 1903), 625.

53. *Literary Digest*, 39 (27 November 1909), 962.

54. *Critic*, 35 (August 1899), 754-756.

55. *Book Buyer*, 22 (March 1901), 148.

56. *Current Literature*, 36 (January 1904), 113.

57. *Scribner's*, 36 (October 1904), 394.

58. *Outlook*, 78 (3 December 1904), 865.

59. "We must ask ourselves before we ask anything else, Is it true?—true

to the motives, the impulses, the principles that shape the life of actual men and women?" (Howells, *Criticism and Fiction and Other Essays*, p. 49).

60. Howells to Samuel L. Clemens, Bethlehem, N.H., 9 August 1885, in *Mark Twain-Howells Letters*, 2:535–536. One has the impression that Howells used the term "analysis" idiosyncratically. He asserted, for example, that Walter Scott "was tediously analytical where the modern novelist is dramatic" (*Criticism and Fiction*, p. 17).

61. New York *Times, Saturday Review of Books and Art*, 10 (27 May 1905), 338. An anonymous reviewer for the *Edinburgh Review*, quoted in *Harper's Weekly* (47 (14 February 1903), 273), had declared that James had "added a new conception of reality to the art of fiction."

62. *Nation*, 78 (4 February 1904), 95.

63. H. G. Dwight, "Henry James—'In His Own Country,'" *Putnam's* 2 (July 1907), 438–439. Dwight's conception of modernism foreshadows critical opinion in succeeding decades, as summarized, for example, in Irving Howe's perceptive introduction to his anthology, *Literary Modernism* (Greenwich, Conn., 1967), pp. 11–40.

64. Quoted in James E. Miller, Jr., "Henry James in Reality," *Critical Inquiry*, 2 (Spring 1976), 589, 591, 602.

65. Renato Poggioli, *The Theory of the Avant-Garde*, trans. Gerald Fitzgerald, Cambridge, Mass., 1968, p. 120.

66. James, *The Question of Our Speech. The Lesson of Balzac. Two Lectures*, Boston, 1905.

67. H. G. Dwight observed in 1907 that James was "the first English writer to reflect certain tendencies of European art" and also "the first American man of letters to be a citizen of the world"—but that his cosmopolitanism "is the thing that his own people most lay up against him. They can forgive almost any of his shortcomings before they can forgive his exile" (*Putnam's*, 2:439).

68. See p. 73 above.

69. John T. Frederick, *The Darkened Sky: Nineteenth-Century American Novelists and Religion*, Notre Dame, Ind., 1969, pp. 234–235; Martha Banta, *Henry James and the Occult: The Great Extension*, Bloomington, Ind., 1972, pp. 3, 8, 65.

70. "Mr. Froude's Short Studies," *Nation* (31 October 1867), reprinted in *Literary Reviews and Essays*, pp. 272–273.

71. Santayana, *The Genteel Tradition*, pp. 85–88. See above, p. 13.

72. Originally published in *Ladies' Home Journal*, 1894; reprinted in *My Literary Passions*, New York, 1895, p. 98. Howells is stating more urbanely a doctrine essentially the same as the one Mrs. Emma D.E.N.

Southworth sets forth in her novel *Retribution* (1849), in a passage quoted above (p. 15).

73. At the end of "No. 44, The Mysterious Stranger," in *Mark Twain's Mysterious Stranger Manuscripts*, p. 405.

74. James, "Mr. and Mrs. Fields," *Atlantic* (July 1915), reprinted in *American Essays*, pp. 261–262. James had summarized the deluded faith earlier to his friend Rhoda Broughton as the belief that "through the long years we have seen civilization grow and the worst become impossible" (Rye, 10 August 1914, in *Letters*, ed. Lubbock, 2:389).

75. London, 21 March 1914, in *Letters*, ed. Lubbock, 2:361.

76. London, 14 February 1915, in *Letters*, ed. Lubbock, 2:446. In his essay on the founding of the *Nation* written in 1915, James called the later nineteenth century the "Age of the Mistake," remarkable for "its good faith," "the depth of its delusion," "the height of its fatuity" (101, 44–45).

Index

(*Note.* Titles of secondary works especially pertinent to the present study are entered under the authors' names.)